Advances in Anatomy Embryology and Cell Biology

Vol. 157

Springer

Berlin
Heidelberg
New York
Barcelona
Hong Kong
London
Milan
Paris
Singapore
Tokyo

Oldřich Štěrba Milan Klima
Bernd Schildger

Embryology of Dolphins

Staging and Ageing of Embryos and Fetuses of Some Cetaceans

With 44 Figures and 24 Tables

Springer

O. Štěrba
Department of Biology and Zoology,
University of Veterinary and Pharmaceutic Sciences,
Palackého 1-3
61242 Brno, Czech Republic

M. Klima
Dr. Senckenbergische Anatomie, Institut III
J.W. Goethe University, Theodor-STern-Kai 7,
60590 Frankfurt am Main, Germany

B. Schildger
Tierpark Dählhölzli, Tierparkweg 1, 3005 Bern, Switzerland

ISSN 0301-5556
ISBN 3-540-67212-5 Springer-Verlag Berlin Heidelberg NewYork

Library of Congress-Cataloging-in-Publication-Data

Die Deutsche Bibliothek - CIP-Einheitsaufnahme
Štěrba, Oldřich:
Embryology of dolphins : staging and ageing of embryos and fetuses of some cetaceans / Oldřich Štěrba ; Milan Klima ; Bernd Schildger. - Berlin ; Heidelberg ; New York ; Barcelona ; Hong Kong ; London ; Milan ; Paris ; Singapore ; Tokyo : Springer, 2000
(Advances in anatomy, embryology, and cell biology, Vol. 157)
ISBN 3-540-67212-5

Springer-Verlag a member of BertelsmannSpringer
Science + Business Media GmbH

Printed in Germany

Production: PRO EDIT GmbH, 69126 Heidelberg, Germany
Printed on acid-free paper - SPIN: 10718100 27/3136wg - 5 4 3 2 1 0

Preface

The rapid development of molecular biology and genetics has led to renewed interest in embryology, comparative embryology, and studies of the relations between ontogeny and phylogeny. In fact, genes have been identified which are involved in the formation of shapes and structures, and it is becoming apparent that their primary morphological expressions are conspicuously similar in different species. The primarily identical shapes do not become diversified until advanced individualization of embryos, and it is here that it is possible to employ the knowledge of comparative embryology, the branch of science engaged in the study of the development and differentiation of tridimensional structures in different animal groups.

However, comparative embryology has been neglected during the past decades, as its development has appeared to have been completed. In our opinion, the decreased interest in comparative embryology has been caused by the fact that often the time factor was not or could not be respected. In fact, in the case of embryos of wild animals even their ontogenetic age and sometimes the duration of intrauterine development are unknown.

Since the individualization of embryos takes place at the time when the epigenetic mechanisms are affected both by the developmental rates of individual organs and organ systems and by the time relations between ontogenetic events, time becomes the decisive factor. However, with the conception of the timing of ontogeny being still vague, we can only perceive certain developmental stages. The causative connections in the dynamics of the developmental events and of the topical relations changing in time remain concealed; papers published on these topics continue to be descriptive, and hence they are of low informative value. Such results are neither proportional to the difficulties in obtaining material nor to the long preparation of thousands of serial sections and their subsequent toilsome study. Finally, the results of such studies are inadequately estimated by the scientific public, and prospects for gaining grants or other financial support are negligible. In this paper we wish to demonstrate new opportunities for comparative embryology, arising by the application of the time factor in ontogeny.

Acknowledgements

We wish to thank Dr. W. F. Perrin, La Jolla, USA, and Dr. J. F. Willgohs, Bergen, Norway, for help in obtaining material. Our thanks are due to Mrs. I. Kirschenbauer, Frankfurt am Main, for her technical assistance, to Mrs. V. Postníková, Grad. Design., Brno, Czech Republic, who did the drawings, to Mr. M. Novotný, Brno, who did the photographs, and to Dr. Ing. R. Obrtel, CSc., Brno, who translated the paper into English. This study was first made possible by a stipend from the DAAD in 1989. Later it was supported by Grant No. 645109 from the Academy of Sciences of the Czech Republic, by financial contributions received from Johann Wolfgang Goethe University in 1993, from the Administration der Dr. Senckenbergischen Stiftung, Frankfurt am Main in 1995, and from the Centre of Morphology (Dr. Senckenbergische Anatomie) of the J.W. Goethe University (Prof. Dr. Winckler) in 1999.

Contents

1 Introduction

The Centre of Morphology (Dr. Senckenbergische Anatomie) has succeeded in collecting a large series of embryos and fetuses of cetaceans, especially dolphins. For the present study, we selected species represented in that collection by the most complete gradual series of embryos and fetuses. These species included *Stenella attenuata* (Gray 1846), *Stenella longirostris* (Gray 1828), *Delphinus delphis* (Linnaeus 1758), and *Phocoena phocoena* (Linnaeus 1758). Like most such materials, the embryos and fetuses are of unknown age, only the date and place of capture (latitude and longitude) of pregnant females being known. Therefore, this material appeared to be very suitable for being treated by our staging and ageing method (Štěrba 1975, 1995). So far, the method has been successfully employed in staging and ageing terrestrial mammals of the orders Insectivora and Rodentia (Štěrba 1977a,b), *Castor* spp. (Pilleri et al. 1985), and Artiodactyla and Microchiroptera (Štěrba 1979, 1990).

The prerequisite for the application of our method (Štěrba 1995) is the knowledge of:

1. The length and weight of the adult and its life span
2. The length and weight of the newborn and its ontogenetic pattern (altricial or precocial)
3. The duration of intrauterine development
4. The period or periods at which mating and calving take place

We shall discuss the four points in succession below.

1.1 Length and Weight of Adults, Life Span

The following data are available on the sizes, body weights, and life span of adults: Nowak and Paradiso (1983) reported, for the whole genus *Stenella*, that its members attained lengths between 1500 and 3500 mm and fairly old ages up to 50 years.

1.1.1 Stenella attenuata

Nishiwaki (1972) states that males attain body lengths of 2000–2100 mm. Bryden et al.(1986) report males of this species attaining sexual maturity at body lengths of 1950–1970 mm, females at 1810–1870 mm. Males attain physical maturity at body lengths of 2050–2100 mm, females at 1900–1950 mm.

The body weight of this species varies, being around 100 kg. According to Martin (1992), males in the eastern Pacific region attain the mean length of 2230 mm, females 2070 mm. The biggest specimen measured was a male 2570 mm long and weighing 119 kg. The oldest individual attained the age of 46 years. Furthermore, the same author states that males attain sexual maturity at 1940 mm of body length at the age of 12 years, females at 1820 mm and 9 years. According to Nowak and Paradiso (1983), males attain sexual maturity at the age of 6–11 years, females at 4.5–8 years.

1.1.2
Stenella longirostris

Bryden et al.(1986) report males becoming sexually mature after having attained body lengths of 1700–1900 mm, females at 1650–1900 mm. Physical maturity is attained by males at 1750–1900 mm of body length, females at 1700–1950 mm. Harrison et al. (1972) relate the physical maturity of females to body lengths upwards of 2100 mm. Martin (1992) states that, in the Eastern Pacific, males attain sexual maturity when 1600–1700 mm long and 6–9 years old, females at the age of 4–6 years. The mean body length of adults is 2000 mm.

1.1.3
Delphinus delphis

According to Nishiwaki (1972), the maximum body length of this species is 2140 mm and males are 100–200 mm longer than females. Both males and females attain sexual maturity at lengths of 1700–1800 mm, at which time they are assumed to be 3–4 years old. Individuals are assumed to live 25–30 years long. Bryden (1972) and Bryden et al. (1986) state that males are sexually mature when having attained 1800 mm in body length, females at 1600 mm. These authors consider males that have attained 1900 mm of body length, and females 1800 mm, to be physically mature. Kleinenberg (1956) concludes that males are sexually mature when 1710–2000 mm long, females when 1610–2000 mm long. Schildger (1986) considers 2100 mm and 82 kg to be the mean adult body length and weight, respectively. Harrison (1969) assumes that sexual maturity is reached around 3 years, Tomlin (1957) believes that it is at the age of 2–3 years. Nowak and Paradiso (1983) report, on the basis of various sources, that sexual maturity is attained at the age of 3–5 years, with life expectancy of 20 years. According to Martin (1992), males attain sexual maturity at 5–12 years at the average body length of 1600–1900 mm. Hui (1979) found that dolphins in populations living in the tropical Pacific attained sexual maturity as early as 7 years old but that there also were individuals still sexually immature at the age of 12–14 years. Also, the results obtained by Perrin and Reilly (1984) suggest that sexual maturity is attained at a rather higher age. Populations in the Atlantic contained individuals still sexually immature at the age of 13 years (Buffrenil and Collet 1983; Collet 1994). Physically mature individuals of this species attain body lengths of 1500–2500 mm, rarely 2600 mm, and body weights of 60–75 kg. Martin (1992) reports mean body length of 1700–2400 mm and weight at around 75 kg. The maximum observed body length of males was 2600 mm, body weight 135 kg, the maximum body length of females was 2300 mm.

1.1.4 Phocoena phocoena

Nowak and Paradiso (1983) state that this species attains sexual maturity at the age of 3–4 years and lives 6–10 years on average, in exceptional cases, as long as 13 years. Martin (1992) reports the assumed size and weight at attaining sexual maturity to be over 3 years at 1400–1500 mm of body length in males, and 3 years on average at 1500–1600 mm in females. According to the same author, the oldest age recorded for this species is 14 years. More particulars have been presented by Kinze (1994): males attain sexual maturity at the age of 2–3 years at a mean body length of 1310 mm (ranging from 1250 to 1350 mm), females at the age of 3–4 years at a mean body length of 1440 mm (1350–1560 mm). According to Nowak and Paradiso (1983), this species attains body length of 1200–2000 mm for males and 1500–1700 mm for females. Kinze (1994) reports average body lengths of 1300–1800 mm, rarely over 2000 mm, body weights ranging from 25 to 90 kg, rarely over 100 kg; females are bigger than males.

1.2 Length and Weight of Newborns, Ontogenetic Pattern

Regarding the environment inhabited by dolphins, only the precocial ontogenetic pattern can be considered. According to Slijper (1966), the weight at birth of newborn dolphins is comparatively high, amounting to 10–15% of that of their mother, the incidence of twins being less than 1%. As a rule, a single young is born in the tail presentation. Tomlin (1940) assessed the body length of newborn dolphins at 52%, Ohsumi (1966) at 50% of that of the mother. Other data in the literature are scarce, considerably vague, and approximate. We give them here for the species under study in the order stated above.

1.2.1 Stenella attenuata

Bryden et al. (1986) assessed the body length of newborns at around 900 mm. The authors also reported that Kasuya et al. (1976) had assessed the mean body length at 900 mm, whereas Perrin et al. (1976) at 820 mm. Schildger (1986) concluded that the mean body length of newborns was 900 mm. The longest fetus measured by this author was 945 mm long. Nishiwaki et al. (1965) reported that in a group of *S. attenuata* captured in 1961, they found 34 fetuses, the longest of which measured 1060 mm. A newborn captured in that group measured 1040 mm in body length. Also, Ohsumi (1966) reported the body length of a newborn *S. attenuata* to be 1050 mm. Nowak and Paradiso (1983) reported the mean body length of 830 mm, Martin (1992) reported 850 mm in newborns in eastern Pacific populations. The body weight of fetuses was assessed only by Schildger (1986); the biggest fetus weighed by him was 6950 g in body weight.

1.2.2
Stenella longirostris

Harrison et al. (1972) reported body lengths of newborns varying from 750 to 900 mm, Bryden et al. (1986) reported 770 mm according to Perrin et al. (1976), and Mead et al. (1980) reported 800 mm. Nowak and Paradiso (1983) and Martin (1992) agree in stating the mean body length of newborns as 770 mm. The longest fetus examined by Schildger (1986) was 805 mm long.

1.2.3
Delphinus delphis

The length at birth of this species has been reported between 800 and 900 mm (Bryden et al. 1986), 750–900 mm (Harrison 1969; Bryden 1972), and 840 mm (Ohsumi 1966). Martin (1992) states the mean body length of newborns as 800–850 mm, Collet (1994) assessed the mean length of newborns in populations of the northeastern Atlantic as 900–1000 mm at a body weight of up to 10 kg. Schildger (1986) examined a fetus measuring 830 mm in length and 4010 g in weight.

1.2.4
Phocoena phocoena

Bryden (1972) reported the body length of newborns between 650 and 900 mm, Watson (1981) the mean length of newborns of 750 mm, Leatherwood et al. (1983) between 700 and 900 mm (half the mother's body length) and 5–7.5 kg in weight. Nowak and Paradiso (1983) reported newborns' lengths of 700–1000 mm and weights of 6–8kg, Martin (1992) lengths of 700–900 mm, Kinze (1994) lengths of 670–900 mm and weights of 5–7kg.

1.3
Summary

From an analysis of literature data on the body lengths of adults and newborns the mean body lengths of newborns can be considered to be: *S. attenuata* 950 mm (820–1060 mm), *S. longirostris* 800 mm (750–900 mm), *D. delphis* 850 mm (750–1000 mm), and *P. phocoena* 800 mm (650–1000 mm).

1.4
Duration of Intrauterine Development

The gestation period of most cetaceans has been assessed by Slijper (1966, 1973) to last 10–12 months. The duration of intrauterine development of the individual species has been reported as follows.

1.4.1
Stenella attenuata, S. longirostris

Yablokov et al.(1972), based on McBride and Kritzler (1951), state the duration of intrauterine development as 9.5 months, Schildger (1986) 11 months, and Nishiwaki et al. (1965) as long as 12 months. Based on various sources, Evans (1987) states 11.5 months for populations in the eastern tropical Pacific but admits 9–10 months as well, and 11.2 months for populations living off the shores of Japan. Harrison (1969) reports 9–10 months of pregnancy for *S. longirostris*, Evans (1987) 9.5–10.7 months for populations of this species in the eastern tropical Pacific.

1.4.2
Delphinus delphis

Asdell (1946) and Harrison et al. (1972) report 276 days of pregnancy in this species. Having considered data of various authors, Harrison (1969) and Yablokov et al. (1972) are inclined to believe that it is 11 months. For populations living in the northern Atlantic and in the Black Sea, Evans (1987) reports 10–11 months; for those in the northern Pacific, 11.5 months.

1.4.3
Phocoena phocoena

In this species, too, estimates of the duration of pregnancy vary from 6 months (or 183 days; Asdell 1946) to 11 months. Harrison (1969) states around 9 months, Yablokov et al. (1972) 9–10 months with a reference to data presented by Guldberg (1894). Evans (1987) states (8)-11 months, Nowak and Paradiso (1983) and Kinze (1994) agree in stating 10–11 months, Martin (1992) states 11 months.

1.4.4
Summary

Data in the literature on the duration of intrauterine development vary from 6 to 12 months, most frequently 9–12 months. We assume that these data have been mostly obtained from observations made at open sea, conclusions having been made from observed matings, findings of embryos and fetuses in captured females, and the occurrence of the first newborns. There is no doubt that such observations are important in learning about the biology of dolphins. However, data on gestation period, obtained in this way, are mere rough estimates. Adult females of the four dolphin species under study do not differ much in size (their body lengths varying between 1600 and 2000 mm, their weights between 70 and 100 kg). The lengths and weights of newborns do not differ considerably, the lengths varying between 650 and 1060 mm, their weights between 5 and 10 kg. Thus, we are justified in concluding that the intrauterine development of the species under study is of the same duration and does not last longer than 10 months.

1.5 Mating and Calving Seasons

We consider it important to define the periods of mating and calving in order to elucidate further facts associated with reproduction, such as problems of monogamy, polygamy, parental care, duration of lactation period, calving intervals, etc. The literature contains numerous data, sometimes contradictory.

1.5.1 Stenella attenuata, S. longirostris

Harrison (1969) writes that very little is known about the reproduction of this polymorphous genus. He reports that in the central Pacific, *Stenella* spp. mate in spring and summer, and that in the same period the testes of males are active. Bryden et al. (1986) quoted Wells (1984) who examined testosterone levels in a captive male *S. longirostris* in Hawaii. The level rose from basal in April until June when it peaked, whereupon it dropped again down to basal in September. The same author also observed an increase in sexual activity in both captive and wild *Stenella* spp. off the Hawaiian islands in late spring and in summer. Nishiwaki et al. (1965) report *S. attenuata* mating in spring and in autumn. In his synthesizing study, Evans (1987) states two mating seasons of *S. attenuata* in the tropical Pacific, viz., March to June and September to November, and two corresponding calving periods from March to June and September to November. The same author reports matings from April to July and calvings from March to June for the eastern form of *S. longirostris*. According to the same author, the white-bellied form is assumed to mate from March to May and August to October, and to calve from February to April and July to September.

1.5.2 Delphinus delphis

Very different data are available on the mating period. Harrison (1969) places it in spring and summer in the Pacific; he found active seminiferous tubules in the testes of males off southern California in late autumn. Harrison et al. (1972) stated that captive individuals in Florida and California mated in April and that male dolphins showed testes enlarged over most of the year. The same authors place the calving period in midwinter, in Californian waters from December to March. Evans (1987), again according to various authors, reports two mating seasons for populations in the northern Pacific, viz., April to June and October to December. According to the same author, calving then takes place from March to April and September to October. Populations living in the northern Atlantic and in the Black Sea have only one mating season, from July to October, and they calve from June to September.

1.5.3 Phocoena phocoena

Mating takes place from July to October (Asdell 1946), in the waters off western Europe in autumn (Harrison 1969). Calving takes place from April to midJune (Asdell 1946), Harrison (1969) places it, in western European waters, in spring and summer. In the Baltic, according to Moehl-Hansen (1954), Bryden (1972) places the calving period in the latter half of May and the first half of June. These data roughly agree with those presented by Evans (1987) who admits a prolongation of the mating season in the northern Atlantic from April until October and the calving period from March until September.

1.5.4 Summary

It is evident from the above data that the mating period is mostly defined very vaguely but, nevertheless, it has been placed in definite parts of the year. *S. attenuata* and *S. longirostris*, inhabiting warm oceans probably reproduce throughout the year, the same as *D. delphis*, in which species all-year-round reproduction has been evidenced. As for *P. phocoena*, the data are relatively uniform. This species, inhabiting seas in the Temperate Zone, mates from spring till autumn and calving has also been observed roughly in that period, which is not at variance with data on the pregnancy period lasting 9 months.

1.6 Calving Intervals, Duration of Lactation

Evans (1987) has also gathered important data on the duration of lactation and on calving intervals, as presented by various authors.

1.6.1 Stenella attenuata

In populations inhabiting the eastern tropical Pacific, duration of the lactation period shows even year-to-year variation, as reported by Perrin and Reilly (1984), and lasts 13, 17, or as long as 26 months. The calving intervals are 2.5, 2.7, or as long as 3.4 years. For the Japanese populations, Kasuya et al. (1974) report lactation periods varying from 26 to 27 months in different years and calving intervals from 3.5.to 3.9 years.

1.6.2
Stenella longirostris

Again, Perrin and Reilly (1984) report lactation periods varying between localities and stocks and lasting 15, 18, and 19 months, and calving intervals of 2.9, 3, and 3.3. years.

1.6.3
Delphinus delphis

The lactation period is reported concurrently as lasting 19 months in northern Pacific, Atlantic, and Black Sea populations. In Pacific populations, the calving intervals are 2.6 years; in the Black Sea and Atlantic ones, 1.3–2.3 years (Evans 1987).

1.6.4
Phocoena phocoena

Evans (1987) reports the lactation period in the northern Atlantic population as lasting 8 months and the calving interval of 1–2 years, according to various authors. These data, too, will necessarily influence considerations of the life span and sexual maturation of the dolphin species under study.

2 Aims of this Study

The aims of this study can be summarized as follows:

1. First of all, to show applicability of the staging and ageing method to all mammals. To point out the advantage of this method in the processing and evaluation of embryological materials, which are many years old in some cases.
2. To show the way of evaluating, by means of this method, undated embryological material on a much higher qualitative level, thus making it possible to compare the results obtained with data on other mammal species and/or groups and facilitating the formulation of evolutionary conclusions.
3. Including the time coordinate in the so far untimed sequence of developmental events and steps can clarify the rate and consecution of successive organogenetic and histogenetic events.
4. To contribute to better knowledge of late ontogeny in the fetal period, a point neglected so far in all mammal groups.
5. To clarify the time course of the reproduction cycle of dolphins by finding out their mating and calving periods.

2 Aims of this Study

3 Material

The study is based on embryos and fetuses of *Stenella attenuata, Stenella longirostris, Delphinus delphis, and Phocoena phocoena,* deposited in the collection of the Zentrum der Morphologie (Dr. Senckenbergische Anatomie) der J.W. Goethe Universität in Frankfurt am Main. The material of the first three species comes from Dr. W. F. Perrin, National Marine Fisheries Service, La Jolla, California, USA. The embryos originated from freshly dead females which had been found entangled in tunny fishing nets in the tropical eastern Pacific. The material of *P. phocoena* comes from Dr. J. F. Willgohs, Zoologisk Museum, University of Bergen, Norway.

In the text to follow, the names of the above species have been abbreviated as: Sa, *Stenella attenuata;* Sl, *Stenella longirostris;* Dd, *Delphinus delphis;* Pp, *Phocoena phocoena.*

The material consisted of the following:

1. *Stenella attenuata,* 90 embryos and fetuses, 21–945 mm in total length (TL). Of these, 15 embryos have been used to prepare serial sections (see Table 1).
2. *Stenella longirostris,* 33 embryos and fetuses, 22–800 mm in TL (see Table 2).
3. *Delphinus delphis,* 29 embryos and fetuses, 24–830 mm in TL. Of these, 9 embryos have been used to prepare serial sections (see Table 3).
4. *Phocoena phocoena,* 15 embryos and fetuses, 25-170 mm in TL, used to prepare serial sections (see Table 4).

The embryos for serial sections were processed following customary histological methods, sectioned at 10 µm into series of microscopical sections and stained with azocarmine and aniline blue (Azan) or hematoxylin and eosin (H&E).

So far, whole embryos and fetuses have been used to describe the development of external morphology of dolphins (Schildger 1986). At the same time, additional somatic dimensions have been measured and the allometric and allochronic relations between certain dimensions have been studied (Schildger 1986; Štěrba et al. 1994); the present paper is a continuation along this line. Serial sections of some of the specimens have been evaluated in several particular communications (Klima 1978, 1990, 1999; Oelschläger et Buhl 1985, 1987; Klima and van Bree 1985, 1990; Lick 1987; Buhl and Oelschläger 1985, 1988; Meyer et al. 1995; Sedmera et al. 1997a,b) and their results have been used in this synthetic study.

Table 1. List of material: *Stenella attenuata* (Sa)

Stage	Embryo number	Identification number	TL (mm)	CRL (mm)	Girth (mm)	Weight (g)	Histo-logical series	Sectional plane
S3	Sa 2	CWO 808	–	–	10	0.085		
	Sa 3	IMC 552	21	8	9	0.035	K 13	Transverse
	Sa 4	RCD 221	21	10	9	0.090		
	Sa 5	RLC 126	22	9	9	0.051	K 16	Transverse
	Sa 6	DBZ 274	25	9	9	0.062	K 12	Sagittal
S4	Sa 7	RCD 230	28	11	10	0.081		
	Sa 8	CWO 801	32	11	10	0.089		
	Sa 9	DJO 15	33	11	10	0.093		
	Sa 10	CWO 752	–	11	11	0.079		
	Sa 12	CWO 813	35	12	11	0.16		
	Sa 14	IMC 146	38	13	12	0.20		
	Sa 15	DBZ 59	38	13	11	0.22		
	Sa 16	DJO 7	39	12	12	0.19		
S5	Sa 11	GMT 61	33	14	15	0.19		
	Sa 13	IWS 046	35	15	15	0.28		
	Sa 17	DPH 005	39	15	15	0.30	K 17	Sagittal
	Sa 18	TMD 234	39	15	13	0.30		
	Sa 19	GEC 097	–	15	13	0.28		
	Sa 20	ILN 036	41	17	17	0.50		
	Sa 21	LLE 009	41	18	17	0.49		
	Sa 22	MLF 002	41	21	18	0.49		
	Sa 23	GMT 080	42	19	17	0.50		
	Sa 25	DLV 058	43	17	15	0.47		
	Sa 26	TJT 030	43	17	15	0.44		
	Sa 27	AXP 206	43	17	14	0.54		
	Sa 28	IMR 029	43	18	16	0.45		
	Sa 29	RJO 105	43	18	17	0.57	K 22	Transverse
	Sa 30	DBH 420	43	19	18	0.50		
	Sa 31	RSG 177	44	19	17	0.57		
S5+	Sa 33	PAT 002	44	19	16	0.56		
	Sa 34	JAY 42	45	20	17	0.62	K 23	Sagittal

Table 1. *Continued*

Stage	Embryo number	Identification number	TL (mm)	CRL (mm)	Girth (mm)	Weight (g)	Histo-logical series	Sectional plane
S6	Sa 24	GLF 105	42	22	19	0.71		
	Sa 32	GMA 173	44	19	19	0.61		
	Sa 35	HDM 001	48	22	20	0.85		
	Sa 36	WCF 139	49	22	22	0.86		
	Sa 37	DBH 119	52	23	20	0.89		
	Sa 38	IMC 399	53	23	19	0.98		
	Sa 39	IMC 571	57	26	23	1.2	K 18	Sagittal
	Sa 40	DIF 030	64	28	24	1.6		
	Sa 41	ADB 035	65	29	24	1.7		
	Sa 42	LCE 014	66	26	23	1.5		
	Sa 43	DLV 055	67	30	23	1.6		
	Sa 44	MSM 066	70	30	25	1.6		
	Sa 45	RCL 054	71	31	25	1.7		
	Sa 46	RJO 156	71	32	28	2.2	K 24	Sagittal
	Sa 47	RJO 203	74	30	–	–		
S7–	Sa 48	IS 038	76	35	–	–	K 14	Transverse
	Sa 49	TMD 125	85	37	34	3.7	K 15	Sagittal
S7	Sa 50	IMC 196	89	43	35	4.7	K 25	Transverse
	Sa 51	AXP 214	91	44	37	5		
	Sa 52	CWO 127	92	47	36	4.1		
	Sa 53	IMC 419	99	50	38	7.3	K 26	Sagittal
	Sa 54	WCF 180	100	44	36	7.2		
	Sa 55	WCF 159	103	46	45	9.2		
	Sa 56	CWO C 8	107	55	43	8.6		
	Sa 57	WCF 178	110	47	43	8.7		
	Sa 58	CWO R10	110	52	44	8.8		
S8	Sa 59	RWC 130	112	57	51	12		
	Sa 60	ADB 103	112	63	53	13.3		
	Sa 61	KdS 103	114	58	52	12.5		
	Sa 62	SGA 038	114	61	52	12.5		
	Sa 63	GAS 035	115	59	57	11.5	K 27	Transverse

Table 1. *Continued*

Stage	Embryo number	Identification number	TL (mm)	CRL (mm)	Girth (mm)	Weight (g)	Histo-logical series	Sectional plane
	Sa 64	SGA 040	116	46	45	17.7		
	Sa 65	WCF 168	120	62	67	18.6		
	Sa 66	JHT 026	133	61	59	16.8		
	Sa 67	ISC 059	134	73	66	21.4		
S9	Sa 68	JWS 066	151	80	66	26.1		
	Sa 69	DPH 003	153	82	77	44.5	K 28	Sagittal
	Sa 70	LR 5	157	85	76	36		
	Sa 71	IAZ 224	165	86	77	36.5		
	Sa 72	ADB 043	167	93	88	44.5		
	Sa 73	MLF 008	187	101	98	64		
	Sa 74	DJF 029	192	101	88	54.1		
	Sa 75	BDJ 105	193	101	85	52.2		
	Sa 76	ISC 026	205	108	89	55.3		
	Sa 77	RWC 382	213	102	108	81		
	Sa 78	REJ 040	215	110	107	82		
	Sa 79	LR 2	225	112	105	93	K 35	Transverse
S12	Sa 80	ISB 020	245	132	124	133		
S10	Sa 81	TBS 199	261	143	120	138		
	Sa 82	ISC 028	295	152	131	178		
	Sa 83	IWI 114	394	200	174	397		
	Sa 84	SWJ 123	468	220	199	693		
	Sa 85	SWJ 125	477	230	198	719		
S11	Sa 86	REL 118	615	300	275	1525		
	Sa 87	RXM 056	660	340	285	1945		
	Sa 88	BGB 024	690	330	297	2105		
	Sa 89	WJS 002	795	400	323	3225		
	Sa 90	RBR 018	945	510	438	6950		

Table 2. List of material: *Stenella longirostris* (Sl)

Stage	Embryo number	Identification number	TL (mm)	CRL (mm)	Girth (mm)	Weight (g)
S3	Sl 1	RWC 101		–	–	0.030
	Sl 2	WKI 008	22	8	7	0.033
	Sl 3	LGP 190	26	9	8	0.027
S4	Sl 4	RWM 317	31	11	11	0.09
	Sl 5	JAZ 12	33	12	13	0.13
	Sl 6	LGP 174	34	12	13	0.12
	Sl 7	CWO 150	38	13	14	0.18
S5	Sl 8	JWP 71	40	15	14	0.29
	Sl 9	RLC 194	52	22	18	0.84
	Sl 10	JAZ 075	58	23	22	1.1
S6	Sl 11	DJF 038	66	27	26	1.65
	Sl 12	RWC 294	72	28	27	2.1
	Sl 13	GLF 113	78	30	30	2.47
	Sl 14	LGP 169	78	35	32	3.5
	Sl 15	RWC 291	79	33	30	3.15
S7	Sl 16	RSG 56	88	39	37	5.2
	Sl 17	TRK 097	91	40	40	6.1
	Sl 18	RCL 184	96	45	41	6.7
	Sl 19	IWS 062	97	45	37	4.9
S8	Sl 20	GLF 111	114	53	49	10.7
	Sl 21	IMC 143	134	65	67	23.9
	Sl 22	RLC 174	136	66	66	19.6
	Sl 23	GLF 091	137	63	59	16.8
	Sl 24	RSG 71	–	67	59	17.1
S9	Sl 25	GLF 108	177	93	88	50
	Sl 26	RCL 042	198	103	110	90
S12	Sl 27	GMA 33	215	107	107	89
S10	Sl 28	WCF 154	244	108	117	113
	Sl 29	PGT 033	320	137	–	–
S11	Sl 30	MWD 176	590	230	–	–
	Sl 31	GAS 018	700	350	–	–
	Sl 32	PGT 041	720	320	–	–
	Sl 33	PGT 037	805	380	–	–

Table 3. List of material: *Delphinus delphis* (Dd)

Stage	Embryo number	Identification number	TL (mm)	CRL (mm)	Girth (mm)	Weight (g)	Histol-ogical series	Sectional plane
S3	Dd 1	RWC 426	24	8	–	0.029	K 55	Transverse
	Dd 2	DAD 146	31	10	10	0.041		
S4	Dd 3	GMA 67	32	13	13	0.12		
	Dd 4	TMD 588	39	13	13	0.16	K 56	Transverse
S5	Dd 5	REL 70	42	19	20	0.52	K 57	Sagittal
	Dd 6	CCM 074	44	20	21	0.68	K 53	Transverse
	Dd 7	TBS 142	46	21	21	0.86	K 58	Sagittal
S6	Dd 8	IEG 457	52	22	20	0.82		
	Dd 9	IHT 075	54	24	21	1.01		
	Dd 10	BDJ 058	68	29	26	1.7		
	Dd 11	TBS 208	69	30	26	1.7		
S7	Dd 12	REA 025	82	34	32	3.3	K 49	Transverse
	Dd 13	TEL	87	84	35	34	3.3	
	Dd 14	RJO 381	101	40	37	5.5	K 50	Sagittal
S8	Dd 15	RJO 445	114	50	50	10.6	K 59	Sagittal
	Dd 16	WCF	89	120	51	66	13.4	
	Dd 17	DBZ 154	131	59	54	11.2	K 51	Transverse
	Dd 18	TBS 148	135	60	67	19.4		
	Dd 19	BDJ 020	144	56	68	22.3		
S9	Dd 20	CWO 651	163	72	77	32.8		
	Dd 21	BDJ 012	164	74	79	36.8		
	Dd 22	RWC 456	194	77	97	55		
S12	Dd 23	TBS 166	237	116	110	91		
S10	Dd 24	BDJ 015	298	128	154	231		
	Dd 25	RKF 058	300	165	193	456		
	Dd 26	SOB 004	320	172	175	394		
	Dd 27	DBH 161	480	210	228	854		
S11	Dd 28	IWP 166	580	240	259	1250		
	Dd 29	RJO 384	830	420	378	4010		

Table 4. List of material: *Phocoena phocoena* (Pp)

Stage	Embryo number	Identification number	TL (mm)	CRL (mm)	Histological series	Sectional plane
S4	1	3011 B	25	10	MK 74	Transverse
	2	3010 B	28	11	MK 71	Sagittal
	3	4308 B	39	14	MK 72	Transverse
S5	4	2807 B	41	19	MK 73	Transverse
	5	1946 B	42	18	MK 70	Sagittal
S6	6	2293 B	51	22	MK 67	Sagittal
	7	1550 B	60	24	MK 64	Sagittal
	8	2071 B	69	28	MK 75	Sagittal
	9	3283 B	70	29	MK 62	Sagittal
S7	10	3085 B	90	36	K 19 (head)	Transverse
	11	3976 B	95	45	MK 61	Sagittal
	12	2294 B	107	46	MK 69 (trunk)	Transverse
					MK 69b (head)	Frontal and Transverse
S8	13	2821 B	126	55	MK 68 (trunk)	Transverse
					MK 68b (head)	Frontal and Transverse
	14	2648 B	134	60	MK 76	Horizontal and Transverse
S9	15	Amsterdam Zoo	167	—	MK 8 (breast and flipper)	Transverse
					MK 9 (dorsum–median)	Transverse
					MK 48 (head)	Transverse
					MK 49 (fluke)	Transverse
					MK 50 (dorsum–craniad)	Transverse
					MK 53 (dorsum–caudad)	Transverse
					MK 55 (trunk–caudad)	Transverse
					MK 59 (trunk–craniad)	Transverse

"B" following the identification number stands for Bergen.

4 Methods

4.1 Staging

As a first step, it was necessary to sort the material and to classify the specimens in successive stages. To do this, we employed our own staging and ageing method (Štěrba 1975, 1995), classifying the embryos and fetuses in the respective ontogenetic stages (Tables 1–4) by their external appearance, selected characteristics, size, and weight.

In species *Stenella attenuata*, *S. longirostris,* and *Delphinus delphis*, Schildger (1986) measured and weighed all embryos and fetuses and obtained all required numerical values (see Tables 1–3, Tables 5–10). We have adhered to the original digital designation (Fig. 1) of the measurements as follows:

- 1. Crown–Rump length (CRL)
- 2. Total length (TL)
- 3. Crown–Genital distance
- 4. Head length
- 5. Rostrum–Nostrils
- 6. Rostrum–Ear
- 7. Rostrum–Flipper
- 8. Rostrum–Eye
- 9. Eye–Mouth corner (gape)
- 10. Eye–Ear
- 11. Eye–Nostrils
- 12. Genitals–Fluke
- 13. Flipper length
- 14. Flipper width
- 15. Fluke length
- 16. Fluke width
- 17. Dorsal fin length
- 18. Dorsal fin height
- 19. Pelvic limb bud length
- 20. Pelvic limb bud width
- 21. Girth (circumference of thorax)
- 22. Weight (W)

However, we have divided the data into several tables. Lists of material (Table 1, *Stenella attenuata*; Table 2, *S. longirostris*; Table 3, *Delphinus delphis*) contain the

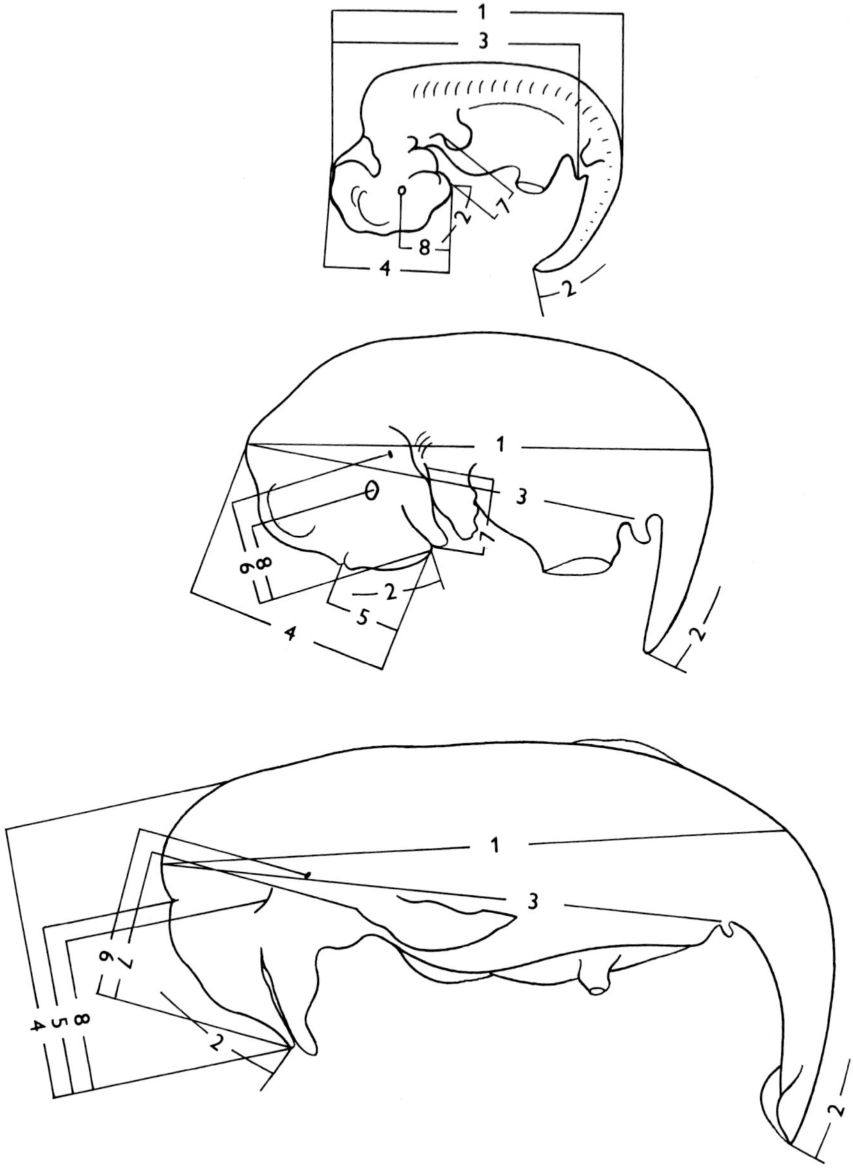

Fig. 1. Illustration of dimensions taken in dolphin embryos and fetuses. *1*, crown–rump length (CRL); *2*, total length (TL); *3*, crown–genital distance; *4*, head length; *5*, rostrum–nostrils; *6*, rostrum–ear; *7*, rostrum–flipper; *8*, rostrum–eye (From Schildger 1986)

measurement values of 2 (TL), 1 (CRL), 21 (Girth), and 22 (W). Table 5 (*S. attenuata*), Table 7 (*S. longirostris*), and Table 9 (*D. delphis*) contain the measurement values of 3–11; Table 6 (*S. attenuata*), Table 8 (*S. longirostris*), and Table 10 (*D. delphis*) those of 12–20.

Table 5. Body measurements in mm (3–11). *Stenella attenuata* (Sa)

Stage	Embryo number	3 Crown–Genitals	4 Head length	5 Rostrum–Nostrils	6 Rostrum–Ear	7 Rostrum–Flipper	8 Rostrum–Eye	9 Eye–Mouth	10 Eye–Ear	11 Eye–Nostrils
S3	Sa 2	–	–	–	–	–	–	–	–	–
	Sa 3	6.0	2.5	0.3	0.6	2.5	0.7	0.5	0.8	0.7
	Sa 4	6.4	3.6	0.4	0.9	2.1	1.0	0.5	1.2	1.0
	Sa 5	6.8	3.3	0.4	0.7	1.7	0.8	0.7	0.8	0.8
	Sa 6	6.2	3.3	0.4	0.8	2.1	1.0	0.7	1.0	0.8
S4	Sa 7	9.0	3.8	0.4	1.0	1.7	0.9	0.7	1.3	1.1
	Sa 8	7.9	4.0	0.5	1.0	2.3	1.7	0.5	1.3	1.4
	Sa 9	8.2	4.2	0.4	1.7	2.5	1.7	0.7	1.6	1.4
	Sa 10	9.9	4.0	0.4	1.0	2.0	1.4	0.5	1.7	1.3
	Sa 12	9.9	5.1	0.4	1.5	2.0	1.4	0.5	1.6	1.7
	Sa 14	11	5.5	0.7	2.0	2.3	1.7	1.3	1.5	1.9
	Sa 15	10.4	5.2	1.2	1.9	2.3	1.5	1.0	1.4	1.9
	Sa 16	10.2	5.0	1.4	1.8	2.2	1.5	1.2	1.7	2.2
S5	Sa 11	12.5	6.6	1.4	2.7	2.3	2.2	0.8	1.7	1.9
	Sa 13	11.8	6.2	1.4	2.5	2.7	2.2	1.2	1.7	2.3
	Sa 17	12.3	6.4	1.0	2.9	2.3	2.2	1.2	1.7	1.8
	Sa 18	12.2	6.5	1.3	2.4	3.4	2.2	1	2	2.2
	Sa 19	–	–	–	–	–	–	–	–	21
	Sa 20	15.1	7.4	2.3	3.9	3.8	3.1	1.7	2.1	3.2

Table.5. *Continued*

Stage	Embryo number	3	4	5	6	7	8	9	10	11
		Crown–Genitals	Head length	Rostrum–Nostrils	Rostrum–Ear	Rostrum –Flipper	Rostrum–Eye	Eye–Mouth	Eye–Ear	Eye–Nostrils
	Sa 21	14.7	7.8	2.2	3	2.8	2.6	1.7	2.2	2.9
	Sa 22	17.3	9.5	2.7	3.4	3.2	3.2	1.4	2.8	4
	Sa 23	15	8.2	2.3	3.3	3	3.1	1.7	2.3	3.6
	Sa 25	13.7	7.8	1	4.2	4.8	3	2	2.1	3.4
	Sa 26	13.6	7.5	1.4	3.3	3.5	3.1	1.6	2.1	2.8
	Sa 27	14.7	8.5	2.2	3.8	3.3	2.4	1.4	2	2.6
	Sa 28	14.3	7.7	1.3	2.9	2.9	2.4	1.3	2	2.3
	Sa 29	15.1	8	2.6	3.1	3	2.6	1.3	2	3.3
	Sa 30	15.4	8.2	2.5	3.6	3.2	2.7	1.9	2.5	3.5
	Sa 31	15.6	8.4	2.3	4.1	3.2	3	1.5	2.6	3.3
	Sa 33	15.3	8.5	2.7	3.6	3.2	2.7	1.4	2.3	3.2
	Sa 34	16.6	8.2	2.6	3.1	3	2.6	1.8	2.4	3.7
S6	Sa 24	18.4	9.3	3	4.1	3.8	3.2	2.2	2.8	4
	Sa 32	15.7	8.7	3.2	3.7	3.2	3.5	2	2.2	3.3
	Sa 35	18.7	10	3.4	4.3	3.6	3.7	2.2	2.3	3.5
	Sa 36	18.2	8.2	3.4	4.6	3.8	4.2	2.2	2.7	4.5
	Sa 37	19.7	10.5	3.7	4.2	3.3	3.5	2.2	2.6	4.4
	Sa 38	19.4	10.4	3.6	4.7	4.3	4	2	2.5	4.7
	Sa 39	21.8	11.4	4.1	5.1	4.2	4.2	2.4	2.7	5.6

Table.5. *Continued*

Stage	Embryo number	3 Crown–Genitals	4 Head length	5 Rostrum–Nostrils	6 Rostrum–Ear	7 Rostrum–Flipper	8 Rostrum–Eye	9 Eye–Mouth	10 Eye–Ear	11 Eye–Nostrils
	Sa 40	24.2	12.9	4.6	6.4	5.8	5.4	3	3.2	5.6
	Sa 41	24.1	12.5	5.2	7	5.4	5.5	2.6	2.8	5.9
	Sa 42	29.2	13	5.2	6.5	5.3	5.8	2.4	2.7	5.3
	Sa 43	25	13.4	5	6.6	5.5	5.9	2.4	2.7	5.5
	Sa 44	24.5	13.7	5.7	8.3	5	6.3	2.4	2.8	5.8
	Sa 45	25.4	14	5.6	8	6.7	7	2.2	2.8	6.3
	Sa 46	27.2	14.2	6	7.9	6.8	7.2	2.5	3.3	6.3
S7	Sa 49	41.3	18.9	8.3	11.7	10.4	9.9	3	3.6	8.2
	Sa 50	41.6	19.6	8.3	12.2	11.3	10.6	3.2	3.4	8.1
	Sa 51	41.5	20.1	9.8	14.3	13.6	11.2	2.7	3.9	8.3
	Sa 52	44.7	19.7	10.1	13.3	12	10.8	2.9	3.6	8.6
	Sa 53	47.3	21.4	10.6	15.2	14.3	13.2	3.7	4.2	10.3
	Sa 54	43	22.3	10	14.8	14.2	12.1	3.2	3.6	9.7
	Sa 55	45.2	24.2	10.4	15.6	15.3	13.3	3.3	4.2	10.5
	Sa 56	50.2	22.9	11.4	16.2	16.3	13.6	3.1	4.2	10.1
	Sa 57	44.9	24.2	11.8	16.2	16	13.7	3.1	4	11
	Sa 58	51	24.5	12.2	16.5	16	14	3.3	4.2	11
S8	Sa 59	56.2	24.4	13.1	17.3	17	14	3.1	4.2	11.6

Table.5. *Continued*

Stage	Embryo number	3	4	5	6	7	8	9	10	11
		Crown–Genitals	Head length	Rostrum–Nostrils	Rostrum–Ear	Rostrum –Flipper	Rostrum–Eye	Eye–Mouth	Eye–Ear	Eye–Nostrils
	Sa 60	62.1	25.1	13.9	19.5	20	15.4	4.3	4.8	12.2
	Sa 61	52.2	25.2	12.2	17.8	17.1	14.5	2.9	4.3	11.7
	Sa 62	51.1	26.7	14.1	18.1	18.8	16.2	3.6	4.8	12.7
	Sa 63	55.2	27	12.8	18.2	19.5	14.5	3	5.2	11.2
	Sa 64	44.5	23.3	12.2	16.5	16.8	14.1	2.3	4.1	10.5
	Sa 65	53.2	28.5	14.8	18.7	21.2	16.7	4	4.9	14.1
	Sa 66	58	30.5	14.4	18.2	20.3	17.5	4	5.3	13.2
	Sa 67	70.6	33	16.2	24.2	27	20	4.8	5.7	18
S9	Sa 68	77	38	20	28	32	23	5	6	16
	Sa 69	80	35	19	28	31	23	5	6	17
	Sa 70	78	37	20	26	28	23	6	7	18
	Sa 71	81	42	22	31	32	24	7	7	19
	Sa 72	90	44	23	32	34	26	7	7	18
	Sa 73	102	47	26	35	37	29	8	8	22
	Sa 74	98	47	26	35	38	29	8	8	19
	Sa 75	99	45	27	33	35	27	8	7	19
	Sa 76	110	47	28	37	40	30	8	8	20
	Sa 77	101	48	29	38	39	31	9	8	22
	Sa 78	108	49	28	38	41	32	8	8	22

Table.5. *Continued*

Stage	Embryo number	3 Crown–Genitals	4 Head length	5 Rostrum–Nostrils	6 Rostrum–Ear	7 Rostrum–Flipper	8 Rostrum–Eye	9 Eye–Mouth	10 Eye–Ear	11 Eye–Nostrils
	Sa 79	111	55	31	40	45	34	8	10	22
S12	Sa 80	130	45	36	46	51	38	10	10	25
S10	Sa 81	138	48	39	49	53	41	10	10	28
	Sa 82	145	54	43	53	58	44	11	11	29
	Sa 83	190	78	62	78	85	64	14	14	40
	Sa 84	205	81	74	81	92	73	15	16	42
	Sa 85	205	90	75	90	96	75	17	17	44
S11	Sa 86	280	118	105	118	110	112	22	18	61
	Sa 87	320	138	119	138	140	119	24	21	61
	Sa 88	280	131	115	131	136	109	23	21	61
	Sa 89	365	160	132	160	178	132	23	31	73
	Sa 90	450	190	165	190	210	145	31	30	77

Table 6. Body measurements in mm (12–20). *Stenella attenuata* (Sa)

Stage	Embryo-number	12	13	14	15	16	17	18	19	20
		Genitals Fluke	Flipper		Fluke		Dorsal fin		Pelvic limb bud	
			Length	Width	Length	Width	Length	Height	Length	Width
S3	Sa 2	4.3	1.4	1.0					0.7	–
	Sa 3	3.2	1.6	0.6					0.6	0.4
	Sa 4	5.6	1.7	0.9					0.5	–
	Sa 5	4.6	1.1	1.0					0.5	–
	Sa 6	4.3	1.2	1.0					0.6	–
S4	Sa 7	5.2	1.4	1.1					0.6	–
	Sa 8	5.4	1.5	1.1					0.6	0.4
	Sa 9	5.8	1.5	1.2					0.5	–
	Sa 10	–	1.4	1.1					–	–
	Sa 12	5.2	1.5	1.4					0.6	–
	Sa 14	6	1.6	1.4					0.5	0.5
	Sa 15	6.2	1.6	1.5					0.7	0.7
	Sa 16	5.4	2	1.4					0.8	0.8
S5	Sa 11	5.3	1.4	1.3					0.2	–
	Sa 13	5	2.1	1.8					0.3	0.3
	Sa 17	6.4	2.1	1.3					0.4	0.4
	Sa 18	7.1	2.2	1.8					0.2	0.3
	Sa 19	6.7	2.3	1.7					0.5	–

Table 6. *Continued*

Stage	Embryo-number	12	13	14	15	16	17	18	19	20
		Genitals–Fluke	Flipper		Fluke		Dorsal fin		Pelvic limb bud	
			Length	Width	Length	Width	Length	Height	Length	Width
	Sa 20	6.8	2.7	1.7	3	1.2			0.1	0.3
	Sa 21	5.8	2.4	2.2					0.3	0.3
	Sa 22	5.5	2.4	2.3					0.4	0.4
	Sa 23	5.3	2.4	2					0.4	0.4
	Sa 25	7.1	3.2	1.8					0.3	–
	Sa 26	5.7	2.3	2.1					0.3	–
	Sa 27	6.3	2.3	2	3.4	1.3			0.1	0.2
	Sa 28	7	2.8	2					0.4	0.4
	Sa 29	6	2.1	2					0.3	0.2
	Sa 30	5.8	2.8	2.2	3.2	1.4			0.1	0.3
	Sa 31	5.1	3	2					0.2	0.2
	Sa 33	5.8	3	2			–	0.3	0.3	0.3
	Sa 34	5.2	2.1	2.2			–	–	0.2	0.2
S6	Sa 24	6.5	3.3	2.1					0.2	0.2
	Sa 32	5.3	3.2	2					0.2	0.2
	Sa 35	6.3	3.4	1.8	4.3	1.6			0.2	0.3
	Sa 36	6.8	3.8	2	–	–			0.2	0.2

Table 6. *Continued*

Stage	Embryo-number	12	13	14	15	16	17	18	19	20
		Genitals–Fluke	Flipper		Fluke		Dorsal fin		Pelvic limb bud	
			Length	Width	Length	Width	Length	Height	Length	Width
	Sa 37	6.3	3.3	2.2	3	1.7			0.3	0.3
	Sa 38	7.7	3.2	2.4	–	–			0.2	0.2
	Sa 39	6.6	3.9	2.3	–	–			0.2	0.2
	Sa 40	9.5	4.2	2.6	5	2.2			0.2	–
	Sa 41	10.6	4.6	2.7	5.2	2			–	–
	Sa 42	10.5	4	2.3	5.2	1.7			–	–
	Sa 43	9.8	4.1	2.9	5.2	2.1			–	–
	Sa 44	12.5	4.3	2.3	5.5	2.3	–	0.2	–	–
	Sa 45	12.6	5.1	2.4	5.1	1.9	6	0.2	0.2	0.2
	Sa 46	11.5	5.2	2.7	5	2.2	6	0.2	–	–
S7	Sa 49	18.3	6.5	3.3	5.9	2.4	9	0.3		
	Sa 50	19.3	6.3	3.7	6.3	3	9	0.5		
	Sa 51	21	7.8	3.3	7.2	4.5	10	0.5		
	Sa 52	21.3	7.6	4.3	7.8	4.1	10	0.5		
	Sa 53	25	9.3	4	7.3	3.5	10	0.5		
	Sa 54	21.5	7.2	4.3	8.5	4.6	10	0.5		
	Sa 55	24.2	8.4	4.2	7.8	4.7	12	0.5		

Table 6. *Continued*

Stage	Embryo-number	12	13	14	15	16	17	18	19	20
		Genitals–Fluke	Flipper		Fluke		Dorsal fin		Pelvic limb bud	
			Length	Width	Length	Width	Length	Height	Length	Width
	Sa 55	24.2	8.4	4.2	7.8	4.7	12	0.5		
	Sa 56	28.2	8	5	7.1	5.4	12	0.5		
	Sa 57	21	8.8	4.3	7	4.3	11	0.5		
	Sa 58	30	8.2	4.5	9	6.5	9	0.7		
S8	Sa 59	27	9.3	4.6	8.8	4.5	10	0.5		
	Sa 60	27.5	9.8	5	9.8	6	11	1.2		
	Sa 61	28.3	9.3	4.7	8.2	5.7	10	0.5		
	Sa 62	27.5	10.4	5.2	9	7.8	12	1		
	Sa 63	31.5	9.9	4.8	10	8.8	14	0.8		
	Sa 64	23	8.3	4.5	9.2	6.5	12	0.5		
	Sa 65	25.8	8.9	5.5	9	10.5	14	2.5		
	Sa 66	35.4	10.3	5.3	10	9.5	14	2		
	Sa 67	34.8	11.7	6.6	11.1	13.7	16	2.3		
S9	Sa 68	39	14	7	13	19	17	3		
	Sa 69	37	14	7	13	17	16	3		
	Sa 70	41	14	7	16	18	16	4		
	Sa 71	45	14	7	14	20	16	5		

Table 6. *Continued*

Stage	Embryo-number	12	13	14	15	16	17	18	19	20
		Genitals–Fluke	Flipper		Fluke		Dorsal fin		Pelvic limb bud	
			Length	Width	Length	Width	Length	Height	Length	Width
	Sa 72	42	15	8	14	23	16	5		
	Sa 73	46	19	9	21	23	20	7		
	Sa 74	47	18	9	17	20	20	6		
	Sa 75	51	17	9	15	30	20	4		
	Sa 76	58	16	9	19	30	23	7		
	Sa 77	49	17	10	20	31	23	8		
	Sa 78	51	21	11	19	34	23	8		
	Sa 79	64	20	11	20	37	24	9		
S12	Sa 80	68	20	12	22	38	30	12		
S10	Sa 81	68	22	12	24	39	30	12		
	Sa 82	69	23	13	23	40	30	11		
	Sa 83	91	31	17	32	63	42	21		
	Sa 84	125	39	21	42	72	55	24		
	Sa 85	129	40	22	40	69	52	22		
S11	Sa 86	161	52	30	55	99	66	35		
	Sa 87	192	72	33	62	97	76	40		
	Sa 88	171	70	35	56	118	92	58		

Table 6. *Continued*

Stage	Embryo-number	12	13	14	15	16	17	18	19	20
		Genitals–Fluke	Flipper		Fluke		Dorsal fin		Pelvic limb bud	
			Length	Width	Length	Width	Length	Height	Length	Width
	Sa 89	235	82	42	76	132	115	59		
	Sa 90	260	102	52	83	157	142	72		

Table 7. Body measurements in mm (3–11). *Stenella longirostris* (Sl)

Stage	Embryo number	3	4	5	6	7	8	9	10	11
		Crown–Genitals	Head length	Rostrum–Nostrils	Rostrum–Ear	Rostrum–Flipper	Rostrum–Eye	Eye–Mouth	Eye–Ear	Eye–Nostrils
S3	Sl 2	6.4	3.1	0.4	0.9	2.1	0.7	0.6	1.1	0.6
	Sl 3	6.3	2.9	0.3	1	2.6	0.7	0.7	1.1	0.7
S4	Sl 4	8.7	4.3	0.5	2.5	2.7	1.7	1.2	1.5	1.6
	Sl 5	10.2	4.9	0.6	2.7	2.8	1.8	1.3	1.7	1.8
	Sl 6	10.5	4.9	0.6	2.7	2.7	1.8	1.3	1.6	1.8
	Sl 7	10.4	5.8	0.5	2.5	2.6	2	1.2	1.7	2.2
S5	Sl 8	12.9	6.5	0.7	2.8	2.7	2	1.5	2.1	2.3
	Sl 9	19.1	9	3.2	5	4.3	4.7	2.7	3.6	4.8
	Sl 10	21.1	10.8	4.2	6.2	5.5	5.4	2.7	2.7	5.2
S6	Sl 11	25.6	14.5	6.2	8	6.6	7.9	2.3	2.6	5.7
	Sl 12	25.2	14.3	6.5	7.7	6.2	6.9	2.9	3.2	6.1
	Sl 13	28.5	16.2	7.4	9.2	8.1	8.2	3	3.2	6.6
	Sl 14	35.4	17.1	7.6	10.1	9.3	9.2	3.1	3.7	6.8
	Sl 15	30.8	16.2	7.2	9.2	9.4	8.2	2.7	3.2	6.9
S7	Sl 16	37.2	19.1	8.8	12	11.1	10.3	3	3.1	7.6
	Sl 17	42.2	20	9.6	10.6	12.1	10.6	3.2	4.3	7.8
	Sl 18	43	22.1	10.8	14.3	15.2	11.6	3.3	3.9	8.5
	Sl 19	43.3	21.8	10.3	14.3	13.3	11.7	3.4	4.1	8.7

Table 7. *Continued*

Stage	Embryo number	3	4	5	6	7	8	9	10	11
		Crown–Genitals	Head length	Rostrum–Nostrils	Rostrum–Ear	Rostrum–Flipper	Rostrum–Eye	Eye–Mouth	Eye–Ear	Eye–Nostrils
S8	Sl 20	51.2	24.8	13	17.7	19.1	15.3	4.5	3.2	11.6
	Sl 19	43.3	21.8	10.3	14.3	13.3	11.7	3.4	4.1	8.7
S8	Sl 20	51.2	24.8	13	17.7	19.1	15.3	4.5	3.2	11.6
	Sl 21	64.2	31.2	17.4	24.4	28.1	20	6.7	5.4	15.6
	Sl 22	62	33.5	17.3	23.5	26.8	20.4	5.2	5	14.7
	Sl 23	60.4	30.2	16.1	20.8	25.5	17.6	5.8	5	14.2
	Sl 24	66	30.2	17.2	23.8	27	19	7	5.8	14.6
S9	Sl 25	90	43	27.2	32.2	37.5	28.4	10.3	6.8	19.7
	Sl 26	105	–	32.2	40.3	50	34.2	11.8	8.9	23.2
S12	Sl 27	108	–	32.2	40.5	49	34.5	12.4	8.1	23.7
S10	Sl 28	109	–	36.5	43.6	48	37.2	13.8	8.8	26.6

Table 8. Body measurements in mm (12–20). *Stenella longirostris* (Sl)

Stage	Embryo number	12	13	14	15	16	17	18	19	20
		Genitals–Fluke	Flipper		Fluke		Dorsal fin		Pelvic limb bud	
			Length	Width	Length	Width	Length	Width	Length	Width
S3	Sl 2	4.7	1.2	1					0.5	0.5
	Sl 3	4.5	1.1	0.9					0.5	0.5
S4	Sl 4	5.5	1.7	1.2					0.5	0.9
	Sl 5	5.6	1.7	1.3					0.9	0.9
	Sl 6	5.7	1.9	1.5					0.8	0.9
	Sl 7	5.7	1.6	1.6					0.7	0.7
S5	Sl 8	5.9	2.1	1.8					0.8	0.8
	Sl 9	5.7	4.1	2.4	1.7	1.6			0.3	0.3
	Sl 10	9.5	4.5	2.3	3.4	1.8			0.5	0.5
S6	Sl 11	11.8	5.3	2.6	4.6	2.1	6	0.2	0.2	0.2
	Sl 12	11.1	6.2	3.2	5	2.1	6	0.2	0.3	0.3
	Sl 13	13.8	6.6	3.4	5.8	2.5	9.8	0.3	0.1	0.1
	Sl 14	13.6	8.3	3.2	6.1	2.4	6	0.3	0.1	0.1
	Sl 15	14.8	7.6	3.1	6	2.7	10	0.5	0.1	0.1
S7	Sl 16	20	8	3.9	6.3	2.6	12.5	0.3		
	Sl 17	20.5	8.1	3.7	6.5	3.7	10.5	0.7		
	Sl 18	22.2	9	4.8	6.8	4.5	12.1	0.8		

Table 8. *Continued*

Stage	Embryo number	12	13	14	15	16	17	18	19	20
		Genitals–Fluke	Flipper		Fluke		Dorsal fin		Pelvic limb bud	
			Length	Width	Length	Width	Length	Width	Length	Width
	Sl 19	23.4	8.6	4	7	4.1	12	0.6		
S8	Sl 20	23.8	11.1	4.8	8.8	8.4	12. 5	0.8		
	Sl 21	32.4	14.4	6.4	10.8	13.8	16	2.2		
	Sl 22	35	14.4	6.8	11.6	12.2	14	2.4		
	Sl 23	34	13.2	6	10.5	11.7	21.2	2.3		
	Sl 24	–	12.1	6.4	–	–	15.3	1.8		
S9	Sl 25	41.5	19.2	8.3	14.7	25.2	21	6.2		
	Sl 26	58	21.7	10	19	31.2	24.3	9.3		
S12	Sl 27	55.2	19.7	10.5	15.2	30	24.2	9.1		
S10	Sl 28	72.3	23.5	10.5	18.2	31.4	21.3	10.9		

Table 9. Body measurements in mm (3–11). *Delphinus delphis* (Dd)

Stage	Embryo number	3 Crown–Genitals	4 Head length	5 Rostrum–Nostrils	6 Rostrum–Ear	7 Rostrum–Flipper	8 Rostrum–Eye	9 Eye–Mouth	10 Eye–Ear	11 Eye–Nostrils
S3	Dd 1	6.5	3	0.4	1.7	2.7	1.4	0.5	1	1.1
	Dd 2	9	3.4	0.6	2.3	2.8	1.6	0.6	1.2	0.8
S4	Dd 3	11.1	5.2	1	2.1	2.1	1.9	1.5	2	1.3
	Dd 4	11.2	5.4	1	2.5	2.4	1.9	1.6	2	1.3
S5	Dd 5	15.5	7.8	3	4.4	4.5	4	2	2.2	3.8
	Dd 6	12.3	8.7	3.2	4.8	5.1	4	2.6	2.6	3.9
	Dd 7	17.6	10	3.5	5.1	5.2	4.4	2.8	3	3
S6	Dd 8	18.2	11.2	4	5.4	5	4.8	2.4	3.1	4.9
	Dd 9	20.6	12.5	4.6	5.9	5	5.5	2.7	2.5	5.1
	Dd 10	24.3	14	6	7.7	6	7	2.3	2.7	6.6
	Dd 11	25.8	14	5.4	11.2	5.6	8.7	2.3	2.8	5
S7	Dd 12	34	17.5	8.4	11.6	10.6	9.7	2.9	3.5	8.6
	Dd 13	32.5	19	8.8	12.8	11.6	9.8	2.9	3.6	8.6
	Dd 14	34.4	20	9.7	16.8	11.9	10.8	3.6	4	9
S8	Dd 15	48	24.5	13.3	17.5	16.8	14.6	4.1	4.6	12
	Dd 16	50.5	25.8	13.8	17.3	17.3	14	4.1	4.8	12.3
	Dd 17	57	26	16	21	15.4	15	4.5	4	12.6
	Dd 18	64	29	18	22	20.5	17.5	6	5	14.7

Table 9. *Continued*

Stage	Embryo number	3 Crown–Genitals	4 Head length	5 Rostrum–Nostrils	6 Rostrum–Ear	7 Rostrum–Flipper	8 Rostrum–Eye	9 Eye–Mouth	10 Eye–Ear	11 Eye–Nostrils
	Dd 19	53.5	31.5	17	28.7	21.5	18.5	5.5	5	14.5
S9	Dd 20	73	35	21.3	28.7	32.3	23.2	7.2	7	17
	Dd 21	73.4	37.4	22.5	28.6	31.4	24.3	7.8	6	17.9
	Dd 22	81	41	26	32.3	35	26.7	8	8.7	20.5
S12	Dd 23	112	52	36	46	49	36	17	10	27
S10	Dd 24	125	67	46	50	63	47	14	11	33
	Dd 25	160	85	65	72	90	67	17	13	42
	Dd 26	192	86	64	73	82	63	16	12	37
	Dd 27	217	102	91	102	110	88	21	17	41
S11	Dd 28	249	105	94	105	111	92	22	16	57
	Dd 29	430	184	153	184	208	153	31	31	83

Table 10. Body measurements in mm (12–20). *Delphinus delphis* (Dd)

Stage	Embryo number	12	13	14	15	16	17	18	19	20
		Genitals–Fluke	Flipper		Fluke		Dorsal fin		Pelvic limb bud	
			Length	Width	Length	Width	Length	Height	Length	Width
S3	Dd 1	3.4	1.5	1.1					0.7	0.7
	Dd 2	5	1.6	1.2					0.7	0.7
S4	Dd 3	5.6	1.8	1.3					0.8	0.5
	Dd 4	6.7	2.2	1.6					1.2	0.7
S5	Dd 5	5.8	3.2	2	2.6	1.2			0.5	0.5
	Dd 6	5.8	3.6	2.1	3.3	1.4			0.5	0.5
	Dd 7	6.5	3.7	2.2	4.1	1.5			0.3	0.3
S6	Dd 8	6.4	4.8	2.3	4.6	1.6			0.3	0.3
	Dd 9	7	4.5	2.3	4.6	1.6			0.2	0.2
	Dd 10	11.6	6.2	2.6	4.8	2.1				
	Dd 11	9.8	6	2.6	4.9	1.8				
S7	Dd 12	17	7.1	3.3	6.6	2.5				
	Dd 13	17	7.1	.3	6.8	3.6		1		
	Dd 14	20.5	8.5	3.7	6.8	3.9		1		
S8	Dd 15	31	9.5	4.7	–	–		1		
	Dd 16	28	8.5	4.7	8.7	7.4	11	1		
	Dd 17	26	10.7	4.5	8.9	9.5	12	1.5		

Table 10. *Continued*

Stage	Embryo number	12	13	14	15	16	17	18	19	20
		Genitals–Fluke	Flipper		Fluke		Dorsal fin		Pelvic limb bud	
			Length	Width	Length	Width	Length	Height	Length	Width
	Dd 18	31	12	6	10.5	12.5	19	3		
	Dd 19	30	12.1	5.3	11	10.5	19	2		
S9	Dd 20	39	14.6	6.8	12.4	14.6	18	4.2		
	Dd 21	41	13.4	7.4	14	19.6	19	4.9		
	Dd 22	44	16.8	7.4	13.2	26.4	19	5.5		
S12	Dd 23	59	24	10	19	37	27	11		
S10	Dd 24	81	26	12	24	48	33	15		
	Dd 25	95	31	15	24	30	46	19		
	Dd 26	91	36	17	27	56	41	21		
	Dd 27	167	45	21	40	91	54	31		
S11	Dd 28	182	52	24	42	85	61	38		
	Dd 29	250	92	40	68	142	105	62		

In *P. phocoena* the total length (TL) and crown–rump length (CRL) were measured just before histological processing and these data are given in Table 4 (according Klima et al. 1985).

In the text to follow, we give the characteristics of stages hitherto used to describe embryonic terrestrial mammals and modified for cetaceans. The corresponding stages in the Carnegie system of staged human embryos (O'Rahilly 1972) are given in brackets.

- **Stage 1.** Primitive streak (Carnegie system S6)
  This stage comprises embryos at the beginning of gastrulation. On the embryonic disc a primitive streak, a primitive pit, and a primitive node can be observed.
- **Stage 2.** First somites (1–7) (Carnegie system S9)
  This stage comprises embryos that begin rising from the flat embryonic disc and change into a cylindrical embryo. Also, the onset of somite formation is observed; this stage comprises embryos showing 1–7 pairs of somites. Embryos showing higher numbers of somites are considered older.
- **Stage 3.** Anterior and posterior limb buds, tail bud, four branchial bars (Carnegie system S13)
  The embryo is spirally twisted and cerebral vesicles protrude at its cranial end. The otic vesicle is closed, the olfactory placode and lens plate are slightly indented. Somite pairs are distinct all along the length of the embryo. Limb buds and tail bud are distinctly developed.
- **Stage 4.** Eye pigmented, handplate present (Carnegie system S16)
  The lens plate is closed to form vesicula lentis. Pigment granules appear in the external layer of the eye cup. Tubercles of the telencephalon are distinctly prominent, the same as the unpaired tubercle of the mesencephalon. Somites are now distinct only in the caudal part of the embryo. Large liver, covered only by mesoderm and ectoderm, is visible in the ventral region of the embryo. The thoracic limb bud is enlarged distally to form the flat handplate.
- **Stage 5.** Handplate indented, pinna present (Carnegie system S19)
  Handplate indentation is not as distinct as in terrestrial mammals. Mesenchymal rays of autopodium can be observed in the handplate but its margin is not indented. In the facial part of the embryo, the nostrils have moved caudally from the growing rostrum. The nasolacrimal groove is closed, and the oral fissure is distinct. Porus acusticus externus lacks pinna. Rudimentary eyelids are present, and the tubercle of the mesencephalon is still visible. The musculature of the lateral wall of the trunk has shifted more ventrally so that the liver is already partly covered.
- **Stage 6.** Palate closed, ossification begins (Carnegie system S23)
  Palatal processes horizontal, fused to form hard palate. First ossification centers appear in the desmal bones of the head. Eyes are partly covered by eyelids. Physiological hernia still persists. This is the last embryonic stage.
- **Stage 7.** Eyelids fused, umbilical hernia reposited
  This is the first fetal stage. All organ rudiments are present, the fetus is separated from the surrounding environment. At this stage the eyelids grow over the eyeball, touching mutually and becoming fused with an epithelial seal, the same as the external acoustic meatus. As the fetus grows it frees a space in the abdominal cavity, and the intestine loops, which have so far reached into a wide umbilical channel, have now been retracted into the abdominal cavity. The musculature of the ab-

dominal wall reaches the median plane ventrally, whereby the umbilical channel is reduced to umbilical cord.

- **Stage 8.** Skin folded or hair follicles present
  These characteristics, typical of terrestrial mammals, are not present in this taxon. In this stage, altricial terrestrial mammals show increasing thickness of the skin, especially corium, which results in the development of numerous skin folds. In precocial terrestrial mammals, the skin increases in thickness and hair follicles appear in it. In cetaceans, no rudiments of hairs or of their auxiliary organs are observed in this stage; only their skin, especially corium, increases in thickness.
- **Stage 9.** Tactile hairs erupted
  At this stage, tactile hairs emerge from the epidermis. These are the only hairs found in cetaceans.
- **Stage 12.** Eyelids open
  At this stage, the epithelial seal connecting the eyelids disappears. In the cetacean species under study, a time shift is observed in that the epithelial seal is freed earlier than in most terrestrial mammals, in which this takes place after having reached stages 10 and 11, and in which the obliteration of the epithelial sealing of eyelids is considered to terminate the fetal period.
- **Stage 10.** First body hairs erupt on the body
- **Stage 11.** Haircoat all over the body
  In terrestrial mammals, these two stages register gradual growth of hairs, from the eruption of the first hairs at stage 10 up to the development of a continuous and dense haircoat at stage 11. In cetaceans only an increase in the amount of skin pigment is observed, becoming apparent at first by light brown coloration. Fetuses showing skin colored various hues of brown are classified in stage 10, those with skin colored black to black-blue in stage 11.

4.2 Ageing

Determining ontogenetic age is the second step of this staging and ageing method (Štěrba 1995). Limiting magnitudes include the body length and body weight of the mother and the length and weight of the newborn (Štěrba 1984). Having analyzed the literary data quoted in the introduction to this paper, we have concluded that adult females of the four species under study do not differ much in size (their total body length varying between 1600 and 2000 mm, their body weight between 80 and 100 kg). It follows from the analysis of data on the body lengths of newborns that the following can be considered the average body lengths of newborns: *S. attenuata*, 950 mm (820–1060 mm); *S. longirostris*, 800 mm (750–900 mm); *D. delphis*, 850 mm (750–1000 mm); and *P. phocoena* 800 mm (650–1000 mm). The body weights of newborns for all four species vary between 5 and 10 kg. Hence, we assume, that the intrauterine development of the species under study is about equally long. On the basis of calculations of the regression curves of allochronic growth of the embryos, it is our opinion that the duration of pregnancy in the dolphin species under study is not longer than 10 months and that the actual duration of pregnancy is closest to 280±10 days, i.e., 270–290 days (see Sect. 5.8).

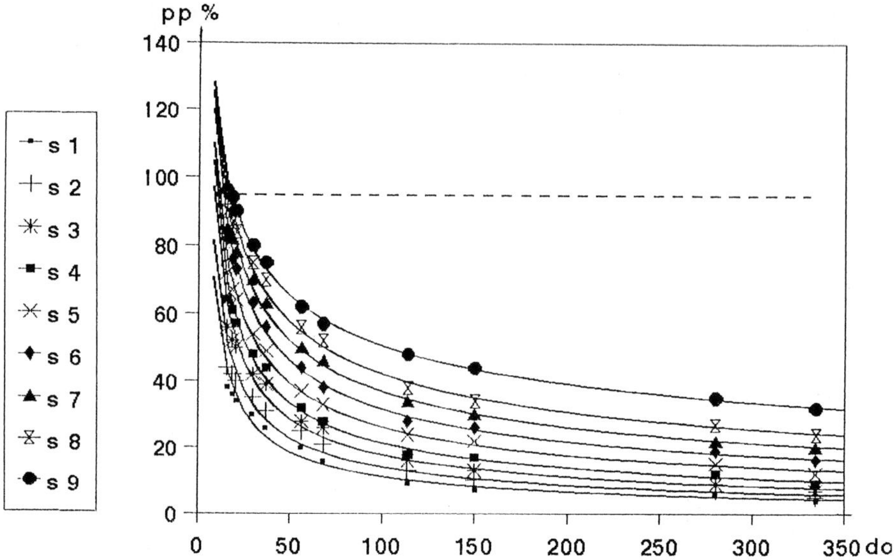

Fig. 2. Regression curves for ageing of staged embryos and fetuses. Ordinate: pp%, percentage duration of prenatal period; abscissa: do, duration of prenatal development in days of ontogenesis. The curves indicate at which percentage of prenatal development the embryos attain ontogenetic stages 1–9. The diagram is valid for mammal species with a gestation period lasting 16–335 days (From Štěrba 1995)

In stages 1–9, the age was read from regression curves calculated for stages 1–9 of mammals whose intrauterine development lasts 16–350 days, as 3rd to 5th degree polynomial regression variances (Štěrba 1995), as shown in Fig. 2 and Table 11.

For the study of relative growth (allometry, standard regression), we used, in agreement with Schildger (1986), the generally used equation of the regression function $y=bx^a$, where y=the size of organ under study, x=the size of the whole organism, b=constant, intercept, and a=the regression coefficient, slope. Plotted in a bilogarithmic coordinate network, the growth follows a regression straight line (lines). The regression coefficient a equals the tangent of the angle between the abscissa and the regression line (tan alpha).

To determine the growth rate of any magnitude in time, we entered, in the above equation, the age on the abscissa in days t, then $y=bt^a$.

This equation illustrates the changes of any organ in relation to time, it informs on the growth rate of the organ or body part in time, on the change in growth rate (acceleration or retardation), and on shifts of the beginning or end of growth into earlier or later periods of intrauterine life.

As a rule, shifts in time have been termed the heterochronies. In the past decades, the term has become a favorite topic of consideration of the evolutional significance of heterochronies and their various types (e.g. Štěrba 1985). In order to prevent incidental misunderstandings and mistakes, we have termed the above changes in

Table 11. Estimated relative age (ERA; %) in developmental stages of embryos and fetuses of dolphin species under study. Presumptive duration of intrauterine development from 270 to 290 days

Stage	270–290 days ERA (%) range	280 days ERA (%) mean	Total length (%) Range	Mean
1	–	4.7	–	–
2	–	6.	–	–
3	8–10	9	1.5–2.5	2
4	10–12	11	2.5–4	3
5	12–15	13	3–5	4
6	15–19	16	4–7	6
7	19–24	20	7–11	9
8	23–28	25	10–14	12
9	29–35	31	15–23	19
12		35		25
10	35–60		30–50	
11	61–100		51–100	
13 (N)		100		100

rates, beginning or end of growth in time, the *allochrony, per analogiam* with the term allometry.

Thereafter we attempted to express the relation of the allometry (1) to the time axis, i.e. allometry in time, $y=bx^a/t$.

The results are shown in Sects. 5.2–5.7. The calculations have been made only in *Stenella attenuata*, of which species the most numerous material was available to us.

In calculations of these relations in stages 3–9 of selected dimensions, we used the mean (x), standard deviation (SD), and standard error of the mean (Sx), see Tables 12–15. In stage 9, the lengths and weights of fetuses began to show considerable differences, so the fetuses were stratified into three groups by 25–50 mm and 10–25 g.

The regression curves for stages 10 and 11 could not be computed for the time being, as characteristics of these stages could be established only vaguely. In most cases they are limited to a single characteristic, as the material available for more precise characteristics was scarce and the lengths and weights of the fetuses show wide variation within the same stage. For this reason, we divided the fetuses into groups stratified by 100 mm of TL. In these groups we then determined the mean ages from the position of the mean TL on the regression growth line from stage 9 to stage 13 (newborn). The data thus obtained were corrected by the weight of fetuses, which in these stages is a more reliable characteristic than the TL or CRL (see also Sects. 5.2–5.7). In all stages, we computed the age of embryos and fetuses for the assumed gestation periods of 270, 280, and 290 days (Tables 16, 17 in *S. attenuata*, Tables 18, 19 in *S. longirostris*, and Tables 20, 21 in *D. delphis*).

From capture dates, we then back-calculated the month in which fertilization had taken place and the month of expected parturition. The results are shown in Sect. 5.9 and in Tables 22–24 and Fig. 32.

5 Results

The results are arranged as follows: After the heading to each stage, a brief characteristic of the stage and estimated age in days are given. Then a description is given of the whole embryo or fetus, its external characteristics and superficial structures. The subsequent paragraphs contain descriptions of the developmental stage of organ systems observed by us and supplemented by earlier data originating in the Zentrum der Morphologie der J.W. Goethe Universität in Frankfurt am Main.

In the concluding paragraph to each stage, the numbers of embryos belonging to the stage in question are recapitulated. If processed to form a histological series, the code number of that series is given in brackets. Then follow data on total length (TL), crown–rump length (CRL) in millimeters, and weight (W) in grams. The development of organ systems has been described up to stage 9, provided that microscopical series were available. In describing stages 10–12, we have confined ourselves to external descriptions only. The ontogenetic development of certain indices and characteristics, such as the growth of the whole embryo and fetus during prenatal ontogeny, growth and changes in the shape of the head, development of the flipper, fluke and dorsal fin are the subject of subsequent paragraphs. In the concluding part of this chapter we shall discuss the assessment of the mating and calving periods.

5.1 Developmental Stages in the Dolphin Species Under Study

5.1.1 Stage 1: Primitive Streak

Estimated age, 13 days. No embryo in this stage has been available. On the basis of calculation of polynomial regression variations, we infer that dolphin embryos attain this stage at the copulation age of 13 days.

5.1.2 Stage 2: First Somites (1–7)

Estimated age, 17 days. Again, none of the embryos examined was in this stage. As in the preceding case, its age has been estimated on the basis of calculations.

5.1.3
Stage 3: Four Branchial Bars, Anterior and Posterior Limb Buds, Tail Bud

Estimated age, range 22–28 days, mean 24 days (Fig. 3; Plates 1, 2).

External Shape. The embryo is spirally twisted, somite pairs are distinct all along its length. The head is approximately spherical. It bears a prominent, high, unpaired bulge of the mesencephalon and, rostral from it, lower, paired pouches of the telencephalon. The otic vesicle is closed and the olfactory placodes are slightly indented. Also, the lens placode is found in various degrees of indentation. Four branchial bars are present in the facial part of head, the first two being visible, the third and fourth submerging or being already submerged in the sinus cervicalis. A cardiac bulge is observed on the ventral side of the embryo. The thoracic limb buds are in the shape of longish and rounded processes arising on the level of the fourth to ninth somites. The pelvic limb buds are considerably smaller than the thoracic ones and they arise on the level of the 34th to 39th somites; they are also rounded distally.

Circulatory System. In this stage, the heart has two atria and one ventricle. Cardiac valves are still not formed, cushions of gelatinous reticulum are present in atrioventricular canals.

Gastropulmonary System. The digestive tube has an epithelial anlage on which the esophagus, stomach, gut, and rudimentary liver can be discerned. The laryngotracheal anlage branches from the ventral wall of the esophageal gut and the two primary bronchi sprout out in a T-like form.

Urogenital System. At this stage, the mesonephros is found in the dorsal part of the abdominal cavity, filling a larger part of the latter.

Nervous System. The brain is in the five-pouched shape, the spinal medulla forming a closed tube. In some embryos the caudal neuropore is still not quite closed.

Hypophysis. Both components that give rise to the hypophysis are visible, viz., the tuber cinereum (the neurohypophysis) and the stomodeal evagination or Rathke's pouch (the adenohypophysis).

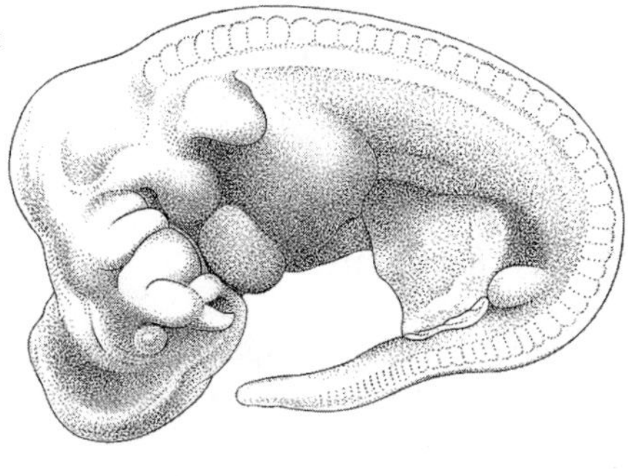

Fig. 3. External appearance of a *Stenella attenuata* embryo *(Sa 5)* in stage 3

Eye. The lens placode is found in various degrees of indentation as well as closed lens vesicle, however, it is not yet separated from the ectoderm. In older embryos the first pigment granules begin to be deposited in the external layer of the optic cup.

Skeletal System. The anlage of the axial skeleton consists of prochondral blastemas of the vertebrae and ribs, which are still fused. The chorda dorsalis passes the anlages of all vertebrae, reaching into that of the occipital bone up to the anlage of the hypophysis. The thoracic limb bud shows axial mesenchymal condensation of the skeletomuscular blastema and a peripheral sinus. In the pelvic limb bud, the mesenchymal condensation is distributed almost uniformly all over the bud.

Muscular System. The blastemas of muscle groups are in the form of mesodermal condensations. The anlage of diaphragma is developed, separating the thoracic and the abdominal cavity.

Integumentum Commune. The epidermis of the trunk consists of a single layer of cubical cells. In the layer of future dermis, thin mesenchyme is present. Lick (1987) observed a similar arrangement of the body integument in Pp.

5.1.3.1 Material

- **Sa:** 2, 3 (K 13), 4, 5 (K 16), 6 (K 12). TL=21–25 mm, CRL=8–10 mm, W=0.035–0.090 g.
- **Sl:** 1, 2, 3. TL=22–26 mm, CRL=8–9 mm, W=0.027–0.033 g.
- **Dd:** 1 (K 55), 2. TL=24–31 mm, CRL=8–10 mm, W=0.029–0.041 g.

5.1.4 Stage 4: Eye Pigmented, Handplate Present

Estimated age, range 27–36 days, mean 30 days (Fig. 4; Plate 3).

External Shape. The embryo is less twisted than in the preceding stage, its somites are now distinctly visible only in the caudal part. A high mesencephalon bulge is still prominent, the same as two small tubercles of the telencephalon rostral from it. The branchial bars 3–4 have gradually sunk into the sinus cervicalis which, in this stage, is hardly or not at all visible. The facial part of the embryo is being formed by gradual fusion of nasal and maxillary processes. The nasolacrimal groove is still visible externally. The olfactory placode is submerged in the nasal pit. The distance between the pits and the rostrum increases in length. The first ectobranchial fissure develops into the meatus acusticus externus whose beginning, the porus acusticus externus, is delimited on all sides in the oldest embryos of this stage. The dark edge of the lens vesicle is conspicuous, caused by the deposition of numerous pigment granules in the outer layer of the optic cup. The thoracic limb bud is distally enlarged to form a flat handplate; in contrast, the pelvic limb bud becomes pointed and gradually diminishes.

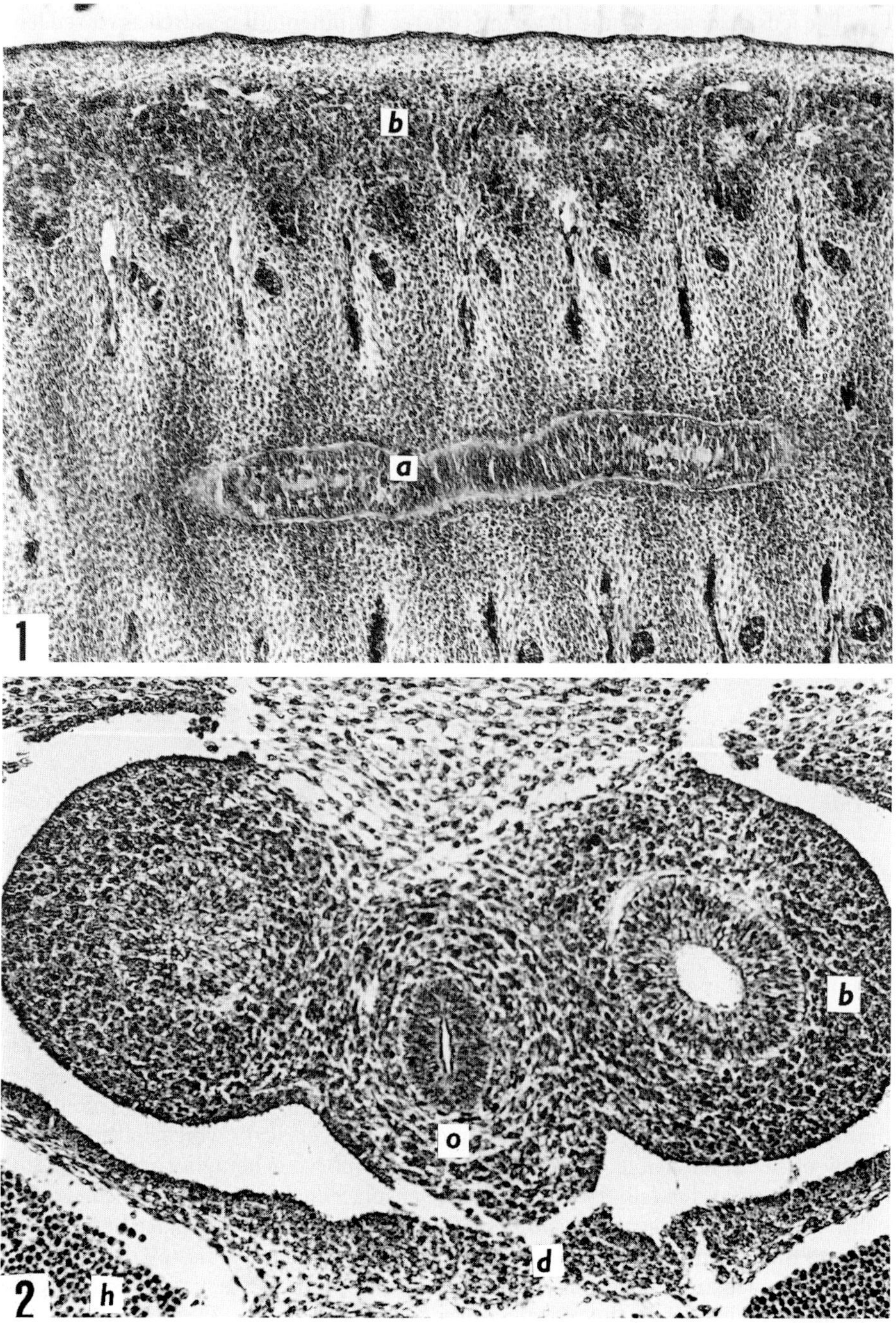

Plate 1. **1** Frontal section through somites; *a*, chorda dorsalis; *b*, myotomes. *D. delphis* (K 55), stage 3, TL 24 mm. Azan, ×100. **2** Cross-section through thorax, caudally of bifurcatio tracheae; *o*, esophagus; *b*, bronchus principalis; *d*, diaphragma; *h*, hepar. *S. attenuata* (K 13), stage 3, TL 21 mm. Azan, ×200.

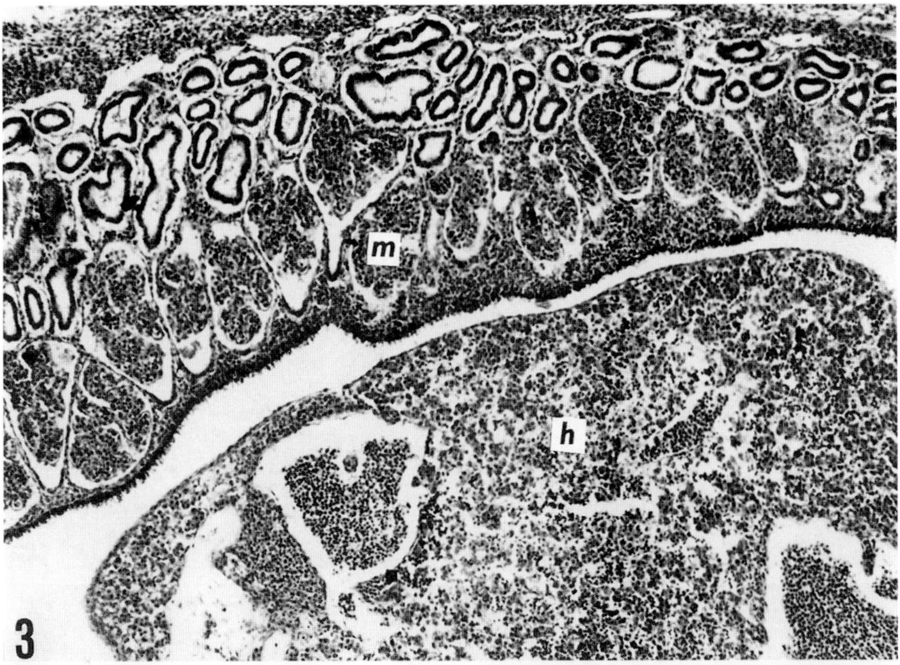

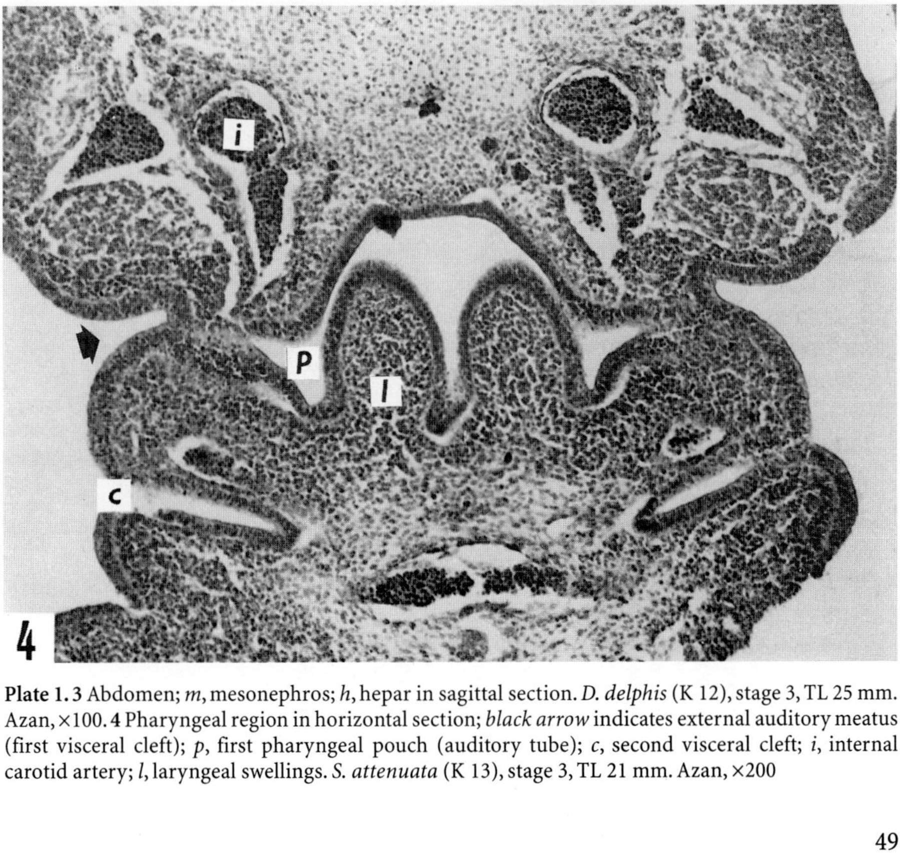

Plate 1. 3 Abdomen; *m*, mesonephros; *h*, hepar in sagittal section. *D. delphis* (K 12), stage 3, TL 25 mm. Azan, ×100. **4** Pharyngeal region in horizontal section; *black arrow* indicates external auditory meatus (first visceral cleft); *p*, first pharyngeal pouch (auditory tube); *c*, second visceral cleft; *i*, internal carotid artery; *l*, laryngeal swellings. *S. attenuata* (K 13), stage 3, TL 21 mm. Azan, ×200

Plate 2. 5 Median section through Rathke's pouch (*white arrow*); *n*, anlage of neurohypophysis; *a*, anlage of adenohypophysis. Chorda dorsalis (*black arrows*) penetrates basisphenoid. *D. delphis* (K 12), stage 3, TL 25 mm. Azan, ×200.
6 Heart in sagittal section; *g*, gelatinous reticulum (endocardial cushions); *v*, ventricle with trabeculae; *h*, hepatic reticulum; *d*, anlage of diaphragma and pericardium. *D. delphis* (K 12), stage 3, TL 25 mm. Azan, ×100.

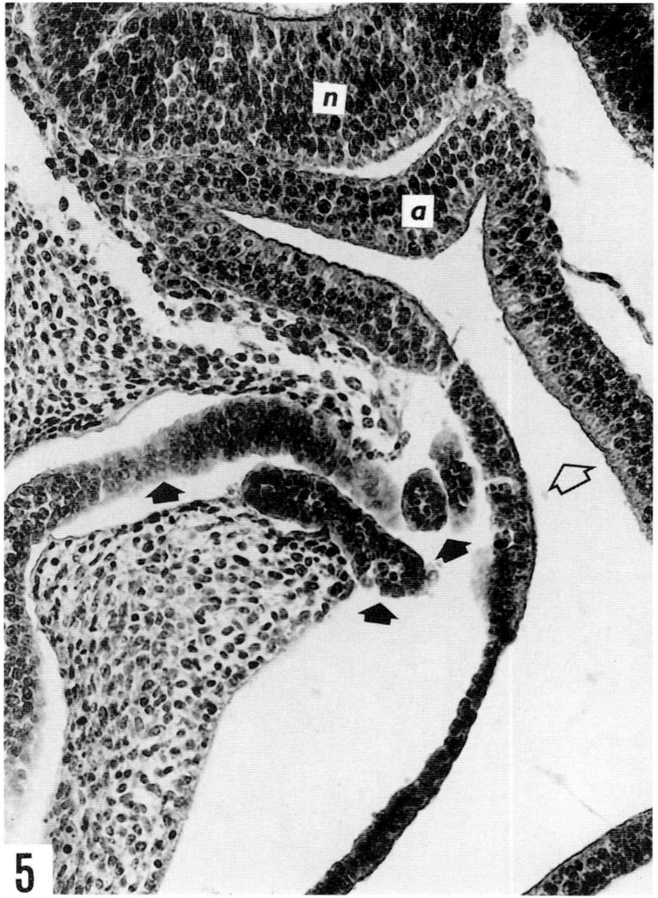

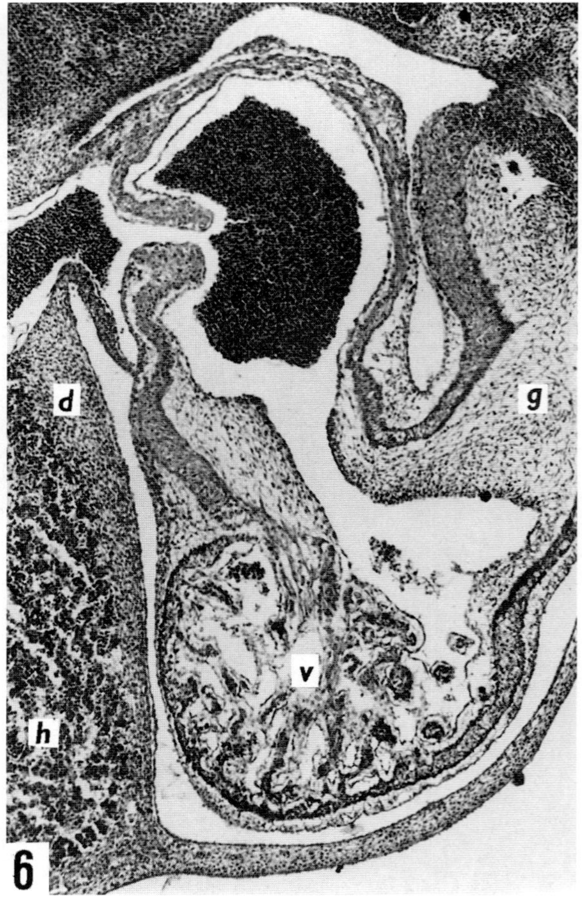

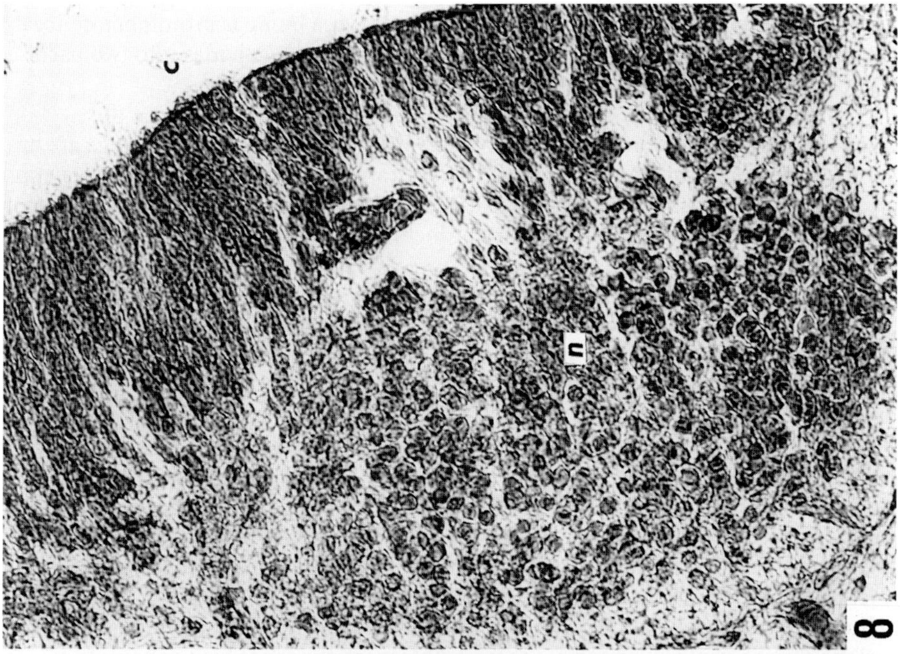

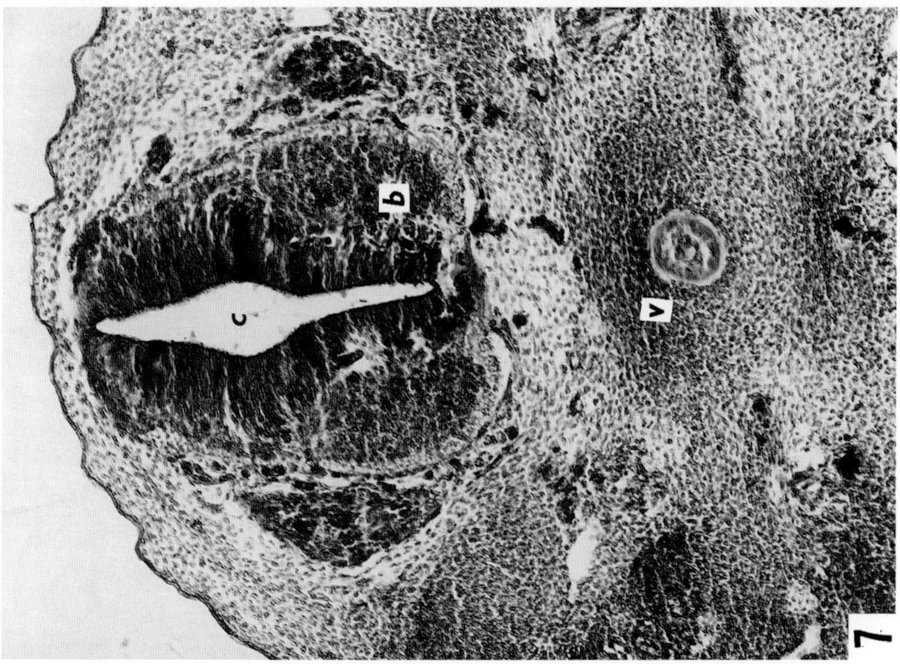

Plate 2. 7 Cross section of medulla spinalis with ganglia spinalia; *c*, canalis centralis; *b*, basal lamina; *v*, condensation in vertebral primordium around notochord. *D. delphis* (K 55), stage 3, TL 24 mm. Azan, ×100. **8** Detail of the previous photograph; *n*, neuroblasts in basal lamina; *c*, central canal. *D. delphis* (K 55), stage 3, TL 24 mm. Azan, ×100

Circulatory System. On the ventral side of the embryo, a bump is prominent, concealing in itself the heart and liver. The heart ventricle gradually divides into two parts: a right and a left one. The chamber wall is as thick as the trabeculae.

Gastropulmonary System. On the labiogingival ridges, epithelial thickenings are apparent as rudiments of dental laminae. The liver anlage, together with the mesonephros, fill most of the abdominal cavity. Near the stomach, the anlagen of both parts of the pancreas are observed. The primary bronchi are developing lobar bronchi.

Urogenital System. The processus uretericus begins to grow into the metanephrogenic tissue, the mesonephros filling the remaining abdominal cavity. The mesonephric (wolffian) ducts open into the cloaca. Gonocytes have moved into the genital ridge and are in an indifferent stage.

Nervous System. The tubercles of the telencephalon are distinctly prominent, while the mesencephalon still forms an unpaired bulge. The brain morphogeny in Pp was studied in detail by Buhl and Oelschläger (1988). On the basis of their own findings in embryonic series of this stage (MK 74, MK 71), they describe the following:
Telencephalon – The anlage of two ovoid telencephalic hemispheres is already distinct. The interhemispheric fissure bifurcates in two telodiencephalic sulci. They delimit the telencephalic hemispheres from the adjacent thalamic and epithalamic regions. On the ventricular surface their counterpart, the torus hemisphericus, forms a prominent ridge which borders the very wide interventricular foramen (Monro). A thickening in the lamina terminalis, rostroventral of the preoptic process, indicates the formation of the commissural plate. The telencephalic hemispheres as a whole are still very small compared to other, already well-developed brain structures.

Diencephalon – The diencephalon is separated from the telencephalon by the torus hemisphericus, the optic recess, and the preoptic recess. The wide and deep infundibular recess, caudally from the optic chiasm, is embraced laterally and ventrally by Rathke's pouch, the anlage of adenohypophysis. The mamillary recess and anlage of the mamillary body demarcates the dorsal end of the diencephalic floor from the tegmentum mesencephali.

Mesencephalon – The midbrain still exhibits the primitive structural pattern of the neural tube. The moderately thickened basal plate, which transforms into the rostral

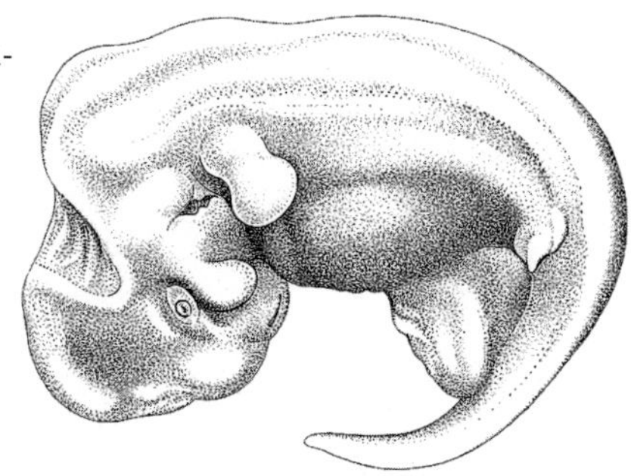

Fig. 4. External appearance of a *Stenella attenuata* embryo *(Sa 16)* in stage 4

tegmentum, is separated by the sulcus limitans from the alar plate that will, in later stages, generate the tectum. This thin-walled mesencephalic tube is strongly curved in the mesencephalic flexure.

Metencephalon – A distinct isthmus rhombencephali constricts the lumen of the neural tube, which opens into the spacious fourth ventricle. The alar plates have given rise to the large cerebellar plates. In the dorsal view they fuse in the midline and form an obtuse angle. In this essentially tubelike brain the cerebellum still exceeds the forebrain in width. It is delimited from the moderately thickened tegmentum metencephali by the pronounced sulcus limitans, which passes through the isthmic channel into the mesencephalic ventricle. The emerging trigeminal nerve is topographically related to the pontine flexure.

Myelencephalon – The myelencephalon is a caudal tubelike part of the brain. The basal and alar plates are separated by the sulcus limitans. In the dorsal view into the wide lumen of the fourth ventricle, the floor is divided by the deep sulcus medianus rhombencephali, which can be traced further rostrally to the mamillary recess. The roof of the ventricle is covered by a transparent epithelial plate. At the cervical flexure, which is approximately 85–90 in both embryos, the myelencephalon passes into the spinal cord with its spacious central canal.

Medulla spinalis – The spinal cord is a closed tube, with neuroblasts proliferating and narrowing the originally spacious central canal to form a dorsoventrally extended fissure with indicated sulci limitantes.

Hypophysis. The stalk of Rathke's pouch begins to narrow.

Eye. The lens vesicle is externally visible as a translucent, hollow sphere separated from the ectoderm. Numerous pigment granules appear in the outer layer of the optic cup, which produce a dark border of the lens vesicle.

Ear. A developing endolymphatic duct can be observed on the capsula otica.

Skeletal System. The nasal skeleton consists of mesenchymal blastemas and young cartilaginous tissue (for Pp, see Klima and van Bree 1985). The prochondral blastemas of vertebrae and ribs are more clearly delimited. The uninterrupted chorda dorsalis passes through all vertebrae. The central mesenchymal condensation in the thoracic limb bud differentiates into individual blastemata, the sinus marginalis is developed. In embryonic Pp, Klima (1990) observed a mesenchymal anlage of the clavicle. The cell condensation in the pelvic limb bud shows no differentiation, the sinus marginalis is indicated.

Muscular System. In the thoracic and abdominal region, a mesodermal ridge separates from the ventral part of fused myotomes. In subsequent stages, this ridge will gradually grow ventrally to form the muscular components of the thoracic and abdominal wall. The abdominal wall consists of only mesodermal and ectodermal layers in this stage. Muscle blastemata are found in the thoracic limb bud.

Integumentum Commune. In all species under study, the epidermis remains on the same developmental level as in the preceding stage and as described by Lick (1987) in

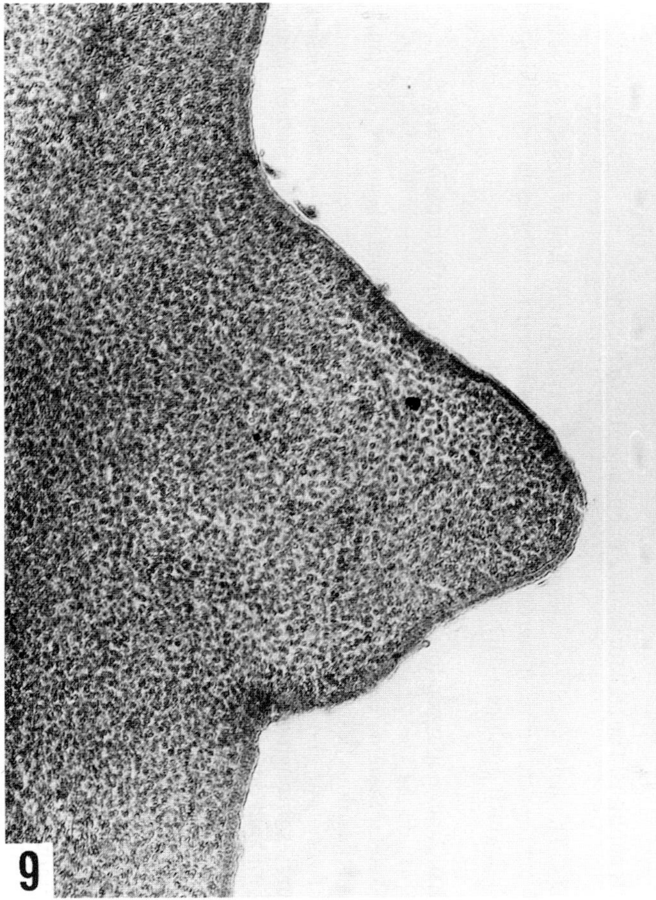

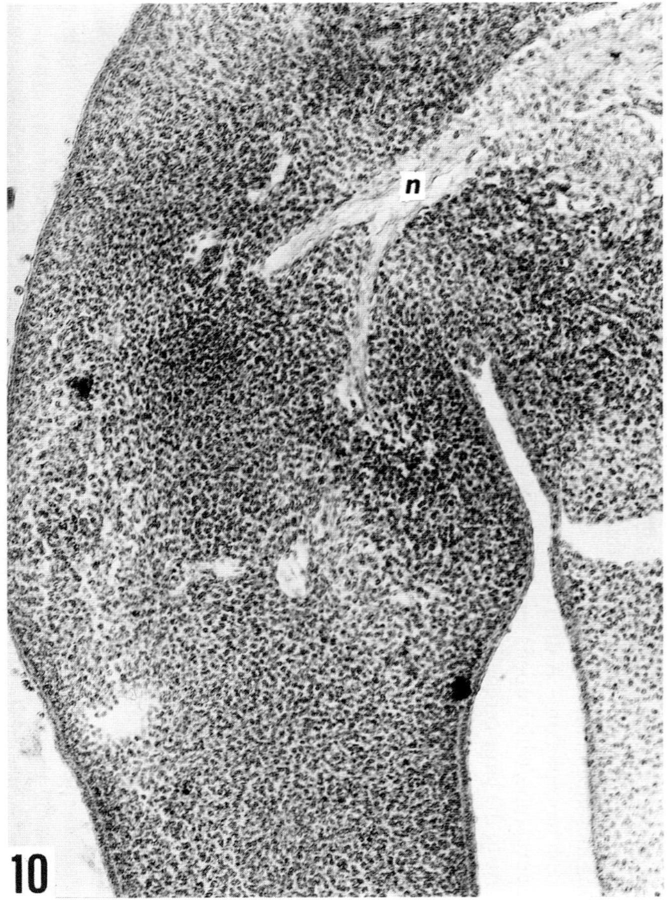

Plate 3. 9 Section through pelvic limb bud. *D. delphis* (K 56), stage 4, TL 39 mm. Azan, ×100. **10** Section through thoracic limb bud with *n*, growing and branching spinal nerve and muscular condensed blastemas. *D. delphis* (K 56), stage 4, TL 39 mm. Azan, ×100.

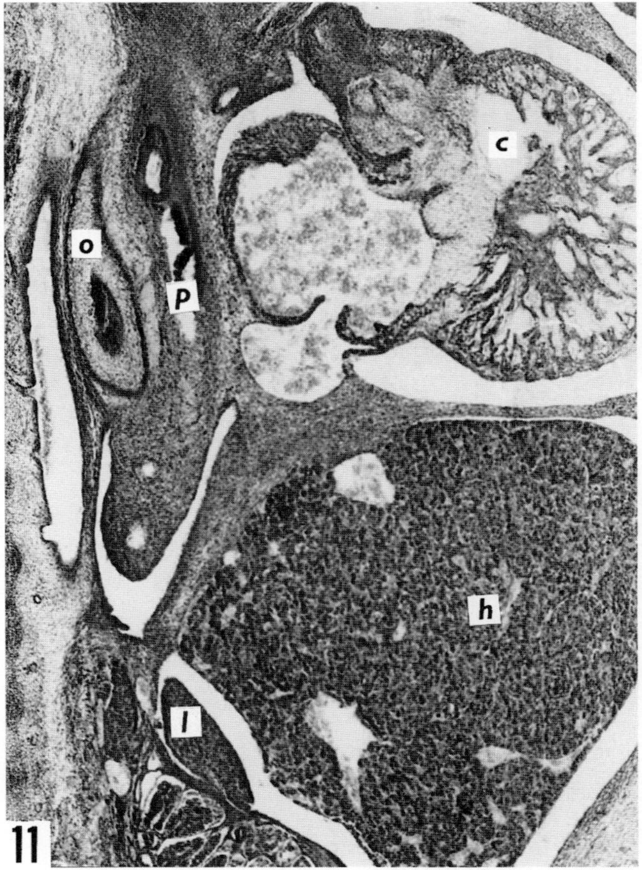

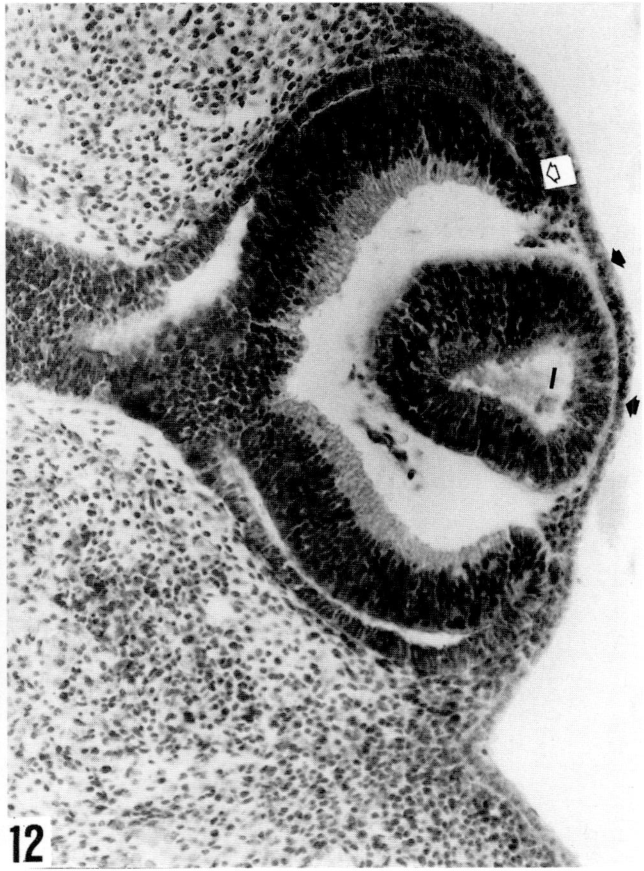

Plate 3. 11 *h*, hepar; *l*, lien; *p*, bronchi and lungs; *o*, esophagus, and *c*, heart in sagittal section. *P. phocoena* (MK 71), stage 4, TL 28 mm. Azan, ×50. **12** Optic cup; *l*, lens vesicle, cornea separated (*black arrows*); pigment in the outer layer from retina (*white arrow*). *P. phocoena* (MK 74), stage 4, TL 25 mm. Azan, ×100

Pp (single layer of cubical cells). Fibroblast differentiation begins in the future corium.

5.1.4.1 Material

- **Sa:** 7, 8, 9, 10, 12, 14, 15, 16; TL=28–39 mm, CRL=11–13 mm, W=0.081–0.22 g.
- **Sl:** 4, 5, 6, 7. TL=31–38 mm, CRL=11–13 mm, W=0.09–0.18 g.
- **Dd:** 3, 4 (K 56). TL=32–39 mm, CRL=13 mm, W=0.12–0.16 g.
- **Pp:** 1 (MK 74), 2 (MK 71), 3 (MK 72). TL=25–39 mm, CRL=10–14 mm.

5.1.5 Stage 5: Handplate Indented, Pinna Present

Estimated age, range 32–42 days, mean 38 days (Fig. 5; Plates 4, 5).

External Shape. In their external form, the embryos differ from those at the preceding stage. The embryonic curvature recedes and the trunk becomes straightened, while the head and the shorter but robust tail remain downturned. The head is spherical with a short facial part already developed, the nasolacrimal groove is closed. The rima oris is distinctly delimited. The nostrils show slight asymmetry: the right nostril is larger. The rostrum increases in length and thus both nostrils are shifted caudally. There is no trace of auricular hillocks, only the porus acusticus externus being present. There are developed anlagen of eyelids. The ventral side of the body is convex, inflated by quickly growing intestines, the intestinal loops being pressed into a wide umbilical sac. In the handplate of the thoracic limb, mesenchymal rays occur, being visible even upon external examination, but the edge of the handplate between them does not become indented as in terrestrial mammals. The handplate is still symmetrical, the axis of the handplate passing the third ray. The pelvic limb buds are pointed and continue to decrease in size. In some of the embryos, slight anlagen of the fluke (Sa 20, Sa 27, Sa 30) and the dorsal fin (Sa 33) have been observed.

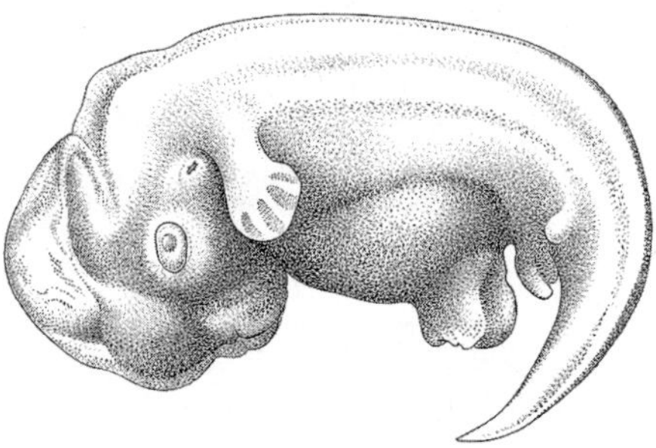

Fig. 5. External appearance of a *Stenella attenuata* embryo (*Sa 33*) in stage 5

Circulatory System. The heart shows two atriums and two ventricles, developing valves, and distinct arterial stems.

Gastropulmonary System. The dental laminae are apparent in both the upper and lower jaws. There is a discernible esophagus, stomach, gut, pancreas, and liver, the latter occupying a large part of the abdominal cavity. Thereby, the lengthening gut is displaced into the wide umbilical stalk, thus giving rise to physiological umbilical hernia. The lungs are divided into lobes with lobal bronchi, the latter being further subdivided into segmental bronchi. The branching of bronchi proceeds.

Urogenital System. Metanephros are present but even mesonephros persists, situated in close proximity, lateral of the gonads. In both sexes, the genital tubercle is found, turned in the caudal direction.

Nervous System. The unpaired bulge of the mesencephalon is still visible in the head. The gray matter increases, the same as the ganglia of the cerebral nerves. The central canal of the spinal medulla is narrowed to form an elongated fissure oriented dorsoventrally, with a wider base.

Hypophysis. The stalk of the adenohypophyseal pouch is closed, epithelial tissue persists in several places. The pouch is transformed into a vesicle with proliferating and folding epithelium.

Eye. The cavity of the lens vesicle is limited to a small fissure.

Skeletal System. According to Klima and van Bree (1985), the nasal skeleton in this stage is formed by young cartilaginous tissue. The vertebrae are cartilaginous, the chorda dorsalis continuously passes through all vertebrae, widening in the intervertebral spaces in cervical, thoracic, and abdominal parts. The ribs are cartilaginous, too. The pectoral limb skeleton – the scapula, humerus, radius, and ulna – is cartilaginous, the anlagen of the autopodium are in a prechondral stage, still not differentiated into the individual segments. As a rule, a desmal rudiment of the clavicle also appears, which can be followed up until stage 8 (Klima 1990). The pelvic limb bud is flattened and transformed into a skin fold.

Muscular System. The myoblasts of the dorsal muscles increase in length and become arranged in the direction of the future muscle fibers. Myoblasts of the muscles of the lateral thoracic and abdominal wall, visible externally as a mesodermal ridge, have moved further in the ventral direction.

Integumentum Commune. The epidermis consists of a basal layer and one or two layers of polygonal peridermal cells (Meyer et al. 1995). Fibroblast differentiation continues in the dermis. Lick (1987) found similar relations in Pp.

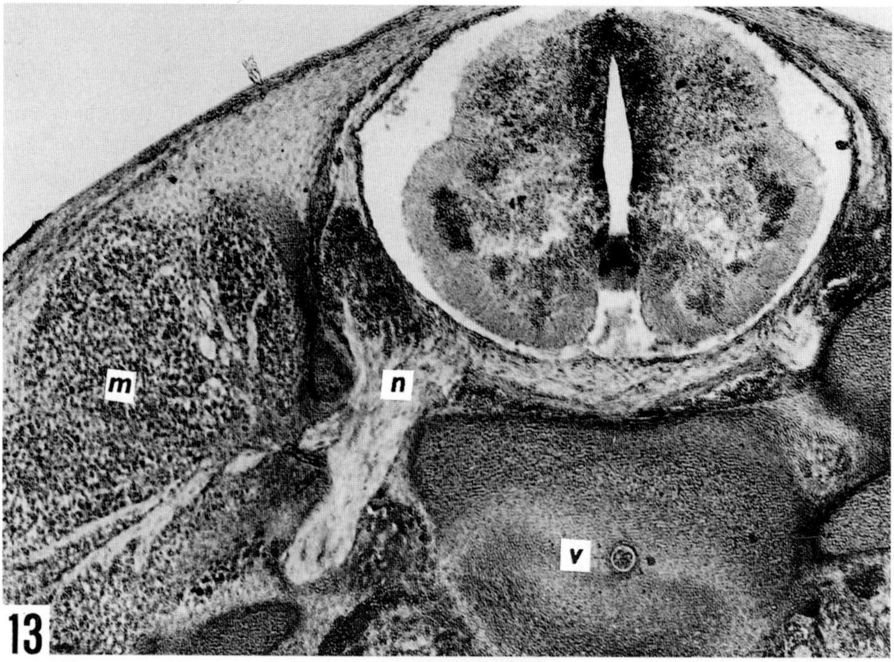

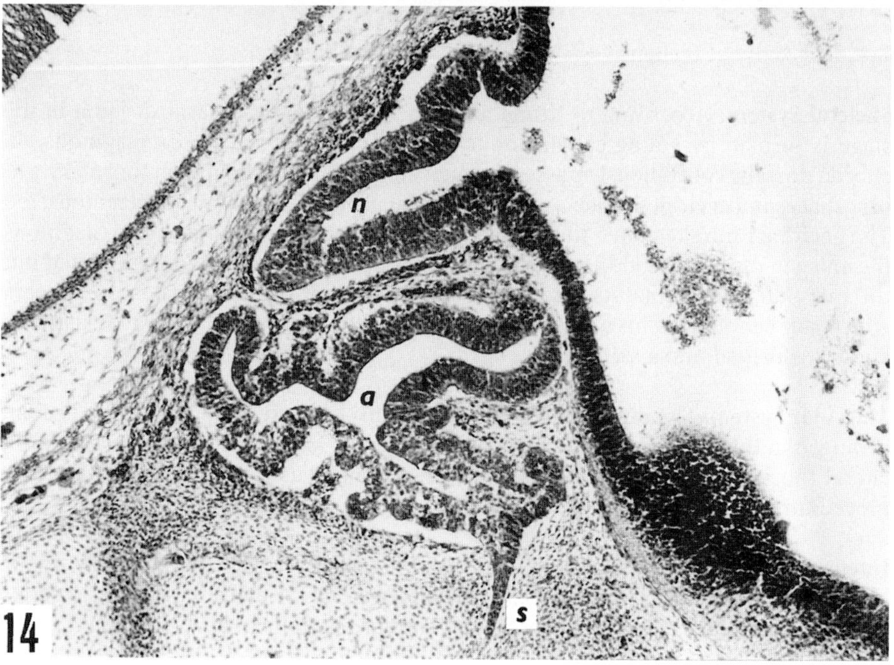

Plate 4. 13 Cross-section through dorsum; spinal cord with central canal in form of an elongated fissure; *n*, spinal nerve leaves spinal canal and divides into dorsal and ventral branches; *m*, primordia of dorsal muscles; *v*, cartilaginous corpus vertebrae with the rest of notochord. *P. phocoena* (MK 73), stage 5, TL 41 mm. Azan, ×40. **14** Primordia of *(n)* neurophysis and *(a)* adenohypophysis; *s*, stalk of Rathke's pouch obliterated. *S. attenuata* (K 23), stage 5, TL 45 mm. H&E, ×100.

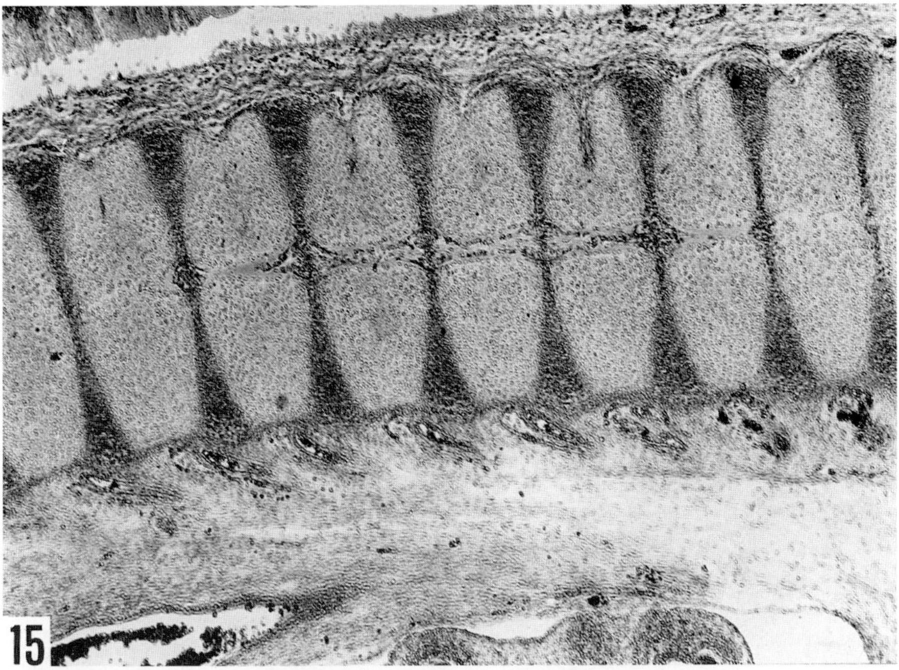

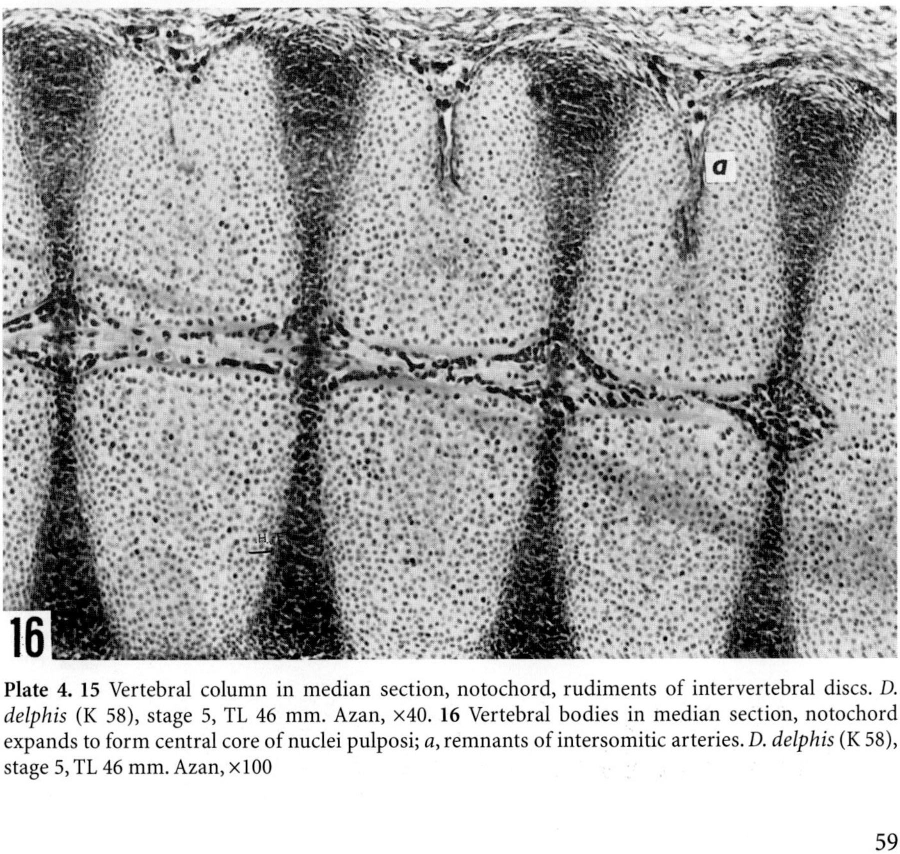

Plate 4. 15 Vertebral column in median section, notochord, rudiments of intervertebral discs. *D. delphis* (K 58), stage 5, TL 46 mm. Azan, ×40. **16** Vertebral bodies in median section, notochord expands to form central core of nuclei pulposi; *a*, remnants of intersomitic arteries. *D. delphis* (K 58), stage 5, TL 46 mm. Azan, ×100

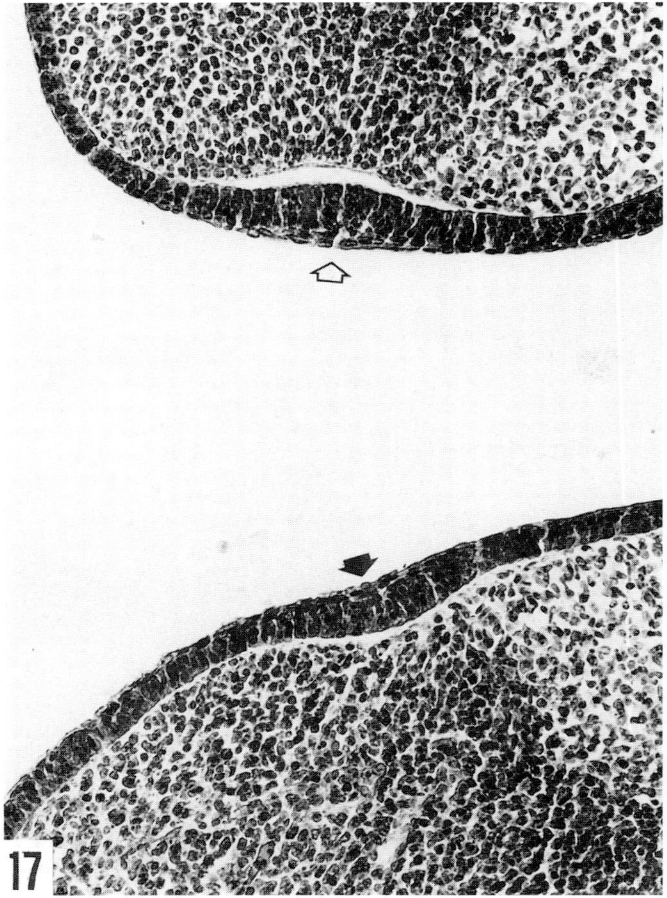

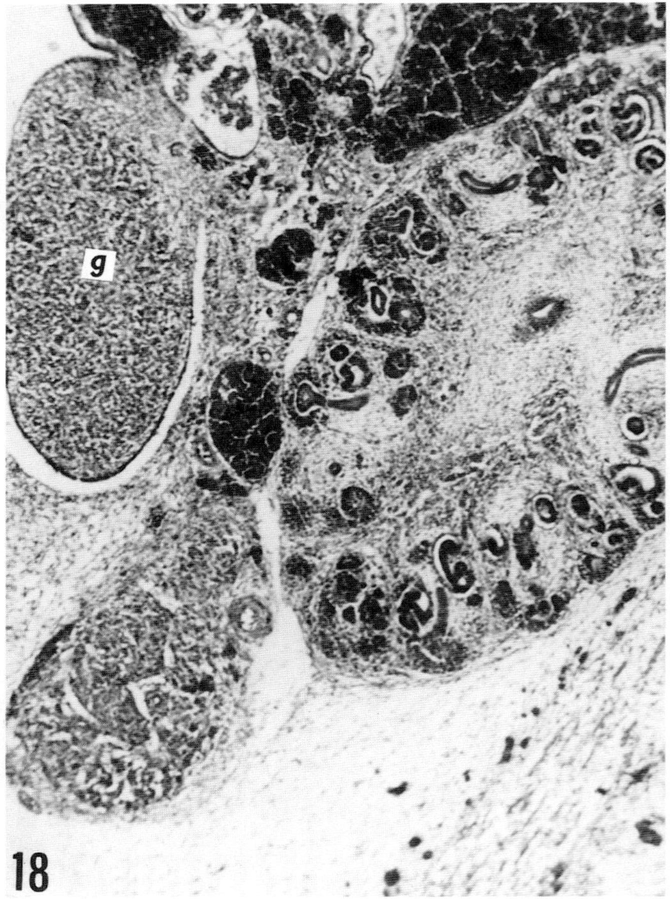

Plate 5. 17 Dental lamina in upper (*white arrow*) and lower (*black arrow*) jaw. *S. attenuata* (K 23), stage 5, TL 45 mm. H&E, ×100. **18** Abdominal cavity with ren and suprarenal gland; *g*, indifferent gonad. *S. attenuata* (K 23), stage 5, TL 45 mm. H&E, ×40.

Plate 5.19 Median section through phallus (*arrow*); *u*, umbilical cord; *v*, caudal vertebrae. *D. delphis* (K 57) stage 5, TL 42 mm. Azan, ×40. **20** Section of intercostal muscles, myoblasts increase in length and arrange up. *D. delphis* (K 57), stage 5, TL 42 mm. Azan, ×250

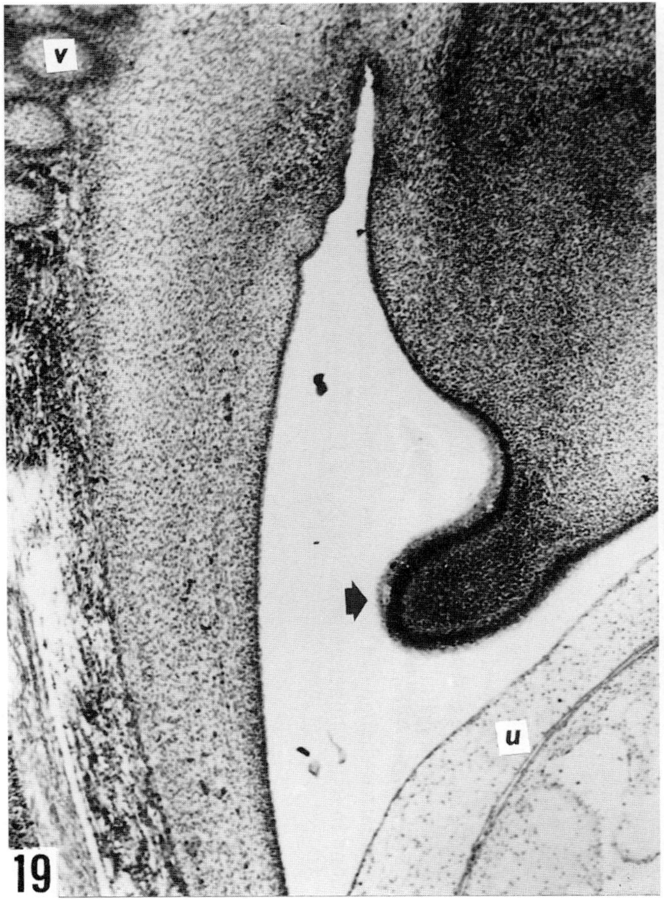

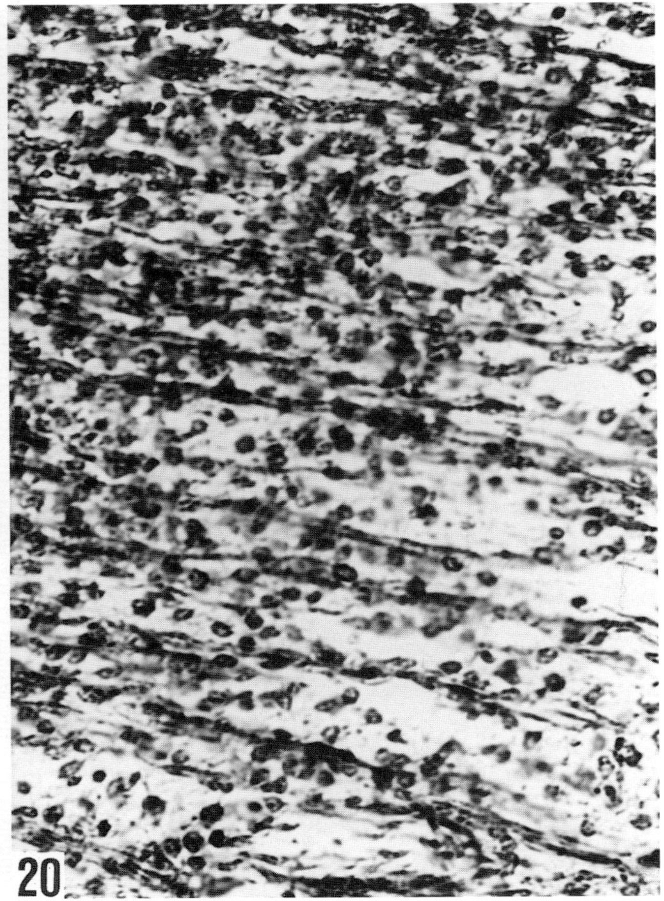

5.1.5.1
Material

- **Sa:** 11, 13, 17 (K 17), 18, 19, 20, 21, 22, 23, 25, 26, 27, 28, 29 (K 22), 30, 31, 33, 34 (K 23). TL=33–45 mm, CRL=14–21 mm, W=0.19–0.62 g.
- **Sl:** 8, 9, 10. TL=40–58 mm, CRL=15–23 mm, W=0.29–1.1 g.
- **Dd:** 5 (K 57), 6 (K 53), 7 (K 58). TL=42–46 mm, CRL=19–21 mm, W=0.52–0.86 g.
- **Pp:** 4 (MK 73), 5 (MK 70). TL=41–42 mm, CRL=18–19 mm.

5.1.6
Stage 6: Palate Fused, Ossification Begins

Estimated age, 41–52 days, mean 46 days (Fig. 6; Plate 6).

This is considered to be the last embryonic stage, as all organs are already developed and a transition to quantitative changes sets in. The fusion of the secondary palate is one of the last qualitative changes that occurs at this stage.

External Shape. The embryo becomes still more straightened in its trunk part and passes into a robust tail in its caudal part. In most embryos, triangular flat folds, the rudimentary fluke, begin to develop on either side of the pointed tail. Also, a low and elongated fold, the anlage of the dorsal fin, appears on the dorsum. The facial part of the head, which is short in the youngest embryos at this stage, begins to turn into the typical beak of the family Delphinidae. The eyes begin to be covered with eyelids. The left nostril, being distinctly smaller in size, becomes closed. It is still open in embryos Sa 24–Sa 36; its opening is indistinct in embryos Sa 37–Sa 43; and from the embryo Sa 44 only one blowhole is observed, situated on the line connecting the eyes, slightly to the right of the median plane. Its seemingly more caudal position is caused by the elongation of the rostral part of the skull. Also, the whole head is slightly asymmetrical. The pectoral limb loses its symmetrical structure, the axis of the handplate is shifted into the second ray, and abduction in the carpal joint takes place, thus starting the external change of the hand into a flipper. Remains of pelvic limb buds are still found in four younger embryos (Sa 24, Sa 32, Sa 35, Sa 36) at this stage on the ventro-

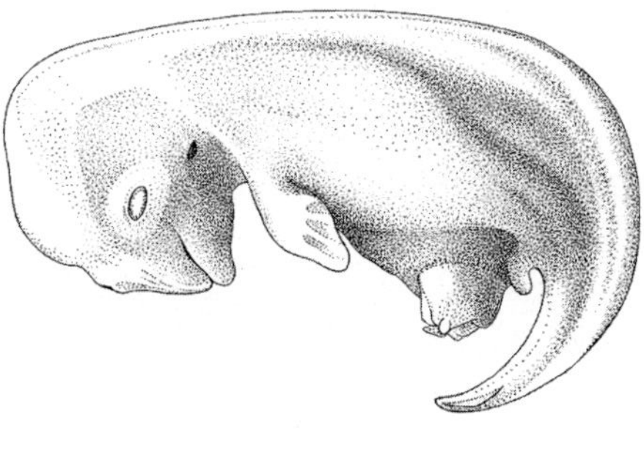

Fig. 6. External appearance of a *Stenella attenuata* embryo (*Sa 40*) in stage 6

caudal surface of the trunk in the form of two minute and pointed tubercles.

Circulatory System. The heart shows developed atrioventricular and semilunar valves. The thickness of the walls is difficult to determine because the angle of section varies; the left ventricle seems to have a thicker wall than the right.

Gastropulmonary System. The palatal processes have assumed a horizontal position and have fused to form the hard palate. The anlage of dentition still exhibits the shape of dental lamina. The abdominal cavity is still wholly filled with the liver and the mesonephros. The wide umbilical stalk containing intestinal loops still persists. The lungs are compact in appearance; sections through the lungs show distinct and numerous bronchial buds. Individual layers begin to differentiate in the wall of the trachea and the large bronchus, and primordia of tracheal rings are particularly distinct.

Urogenital System. The metanephros shows rapid development but a remainder of the mesonephros is still discernible. The gonads differentiate into the testes and the ovaries. The phallus is distinctly prominent in both sexes and its tip is still pointing caudally.

Nervous System. At this stage, Buhl and Oelschläger (1988) observed the degree of development of the encephalon in Pp, series MK 64. The following is a quotation from their paper:

Telencephalon – The growth of the telencephalic hemispheres has markedly proceeded in the dorsal and caudal direction, thereby transforming the telodiencephalic sulci to deep, narrow clefts. A lateroventral expansion of the caudal part of the hemisphere indicates the formation of the lateral pole. In the lateral aspect the outgrowing hemispheres already cover most of the diencephalon. Rostrally they approach each other in the midline and their medial surfaces flatten. In the ventral view the telencephalon already equals the cerebellum in width. Outgrowing fibers and migrating neuroblasts from the olfactory placode have formed the olfactory bulb anlage at the rostroventral extremity of the telencephalon. Both the primordia are located close to the midline and point in a caudal direction. The interventricular foramen (Monro) has considerably narrowed due to expansion of the medial striatal body, thalamus and commissural plate.

Diencephalon – In the lateral aspect most of the thalamic and epithalamic regions are covered by the telencephalic hemispheres. The mamillary tubercle is flattened out in the diencephalic floor. The former wide lumen of the infundibular recess is now obliterated. The thickness of the diencephalic walls has increased and subsequently the lumen of the third ventricle has become considerably narrowed.

Mesencephalon – The mesencephalon has maintained its tubelike appearance, but the expanding diencephalon has shifted the midbrain into a more horizontal position, being situated now above the mesencephalic flexure. The angle of the flexure is compressed to a deep, narrow slit and represents one of several indications for the incipient process of brain foreshortening.

Metencephalon – The angle of the pontine flexure has markedly decreased to less than 90°. The cerebellar anlage has extended caudally, whereas the myelencephalon

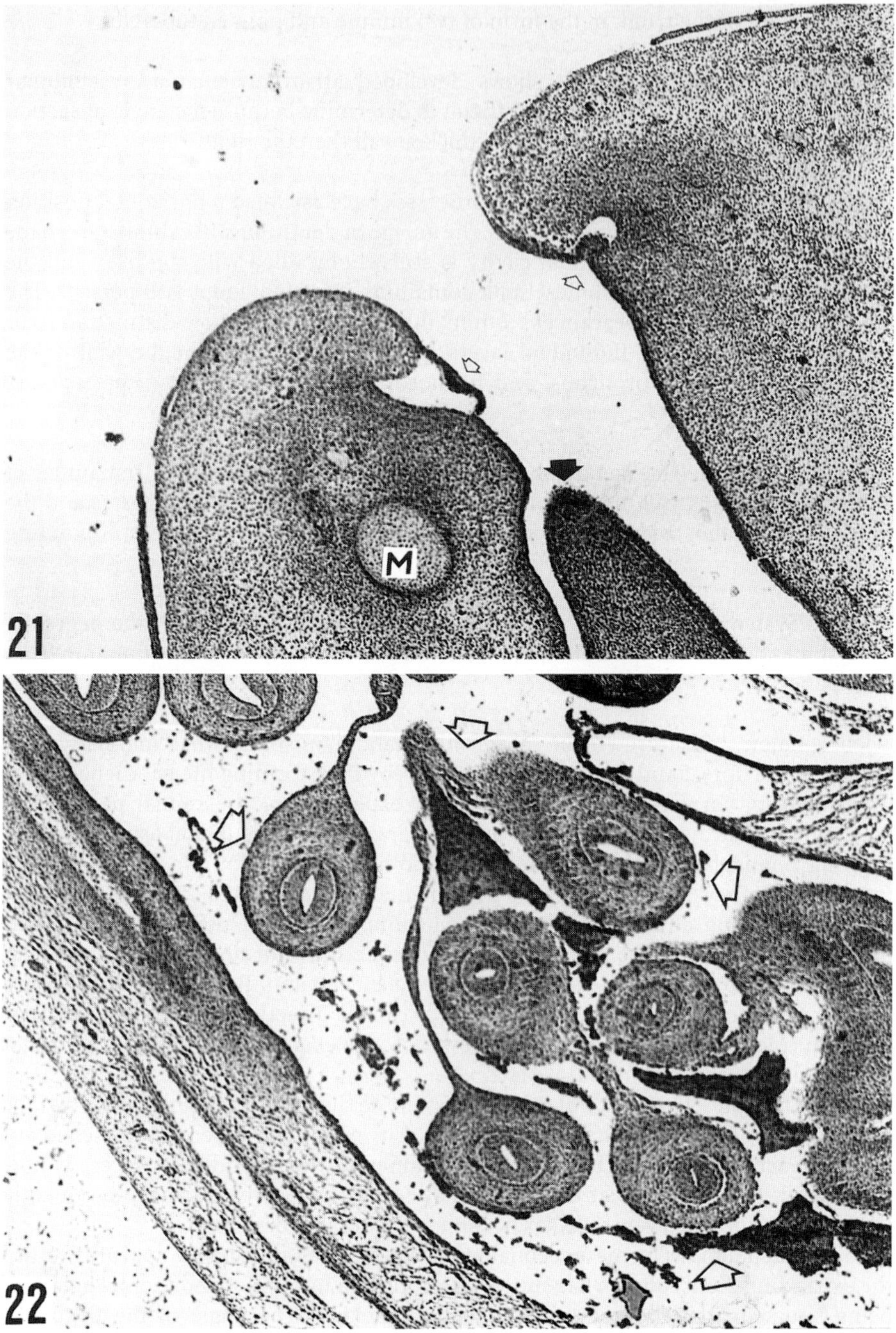

Plate 6. 21 Sagittal section of rostrum; dental buds (*white arrows*); *M*, Meckel's cartilage; apex linguae (*black arrow*). *P. phocoena* (MK 67), stage 6, TL 51 mm. Azan, ×40. **22** Intestinal loops (*white arrows*) in umbilical canal. *P. phocoena* (MK 64), stage 6, TL 60 mm. Azan, ×40.

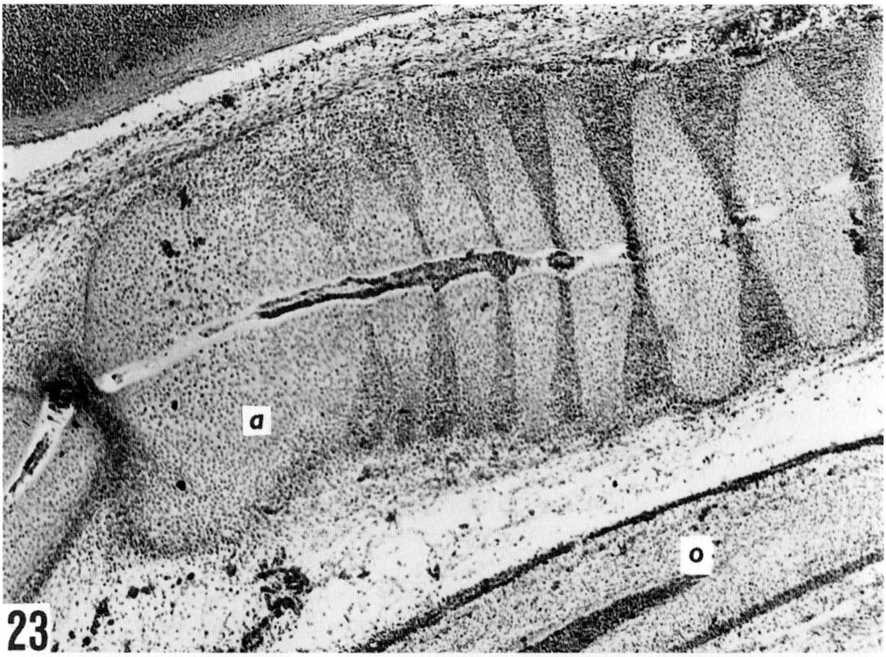

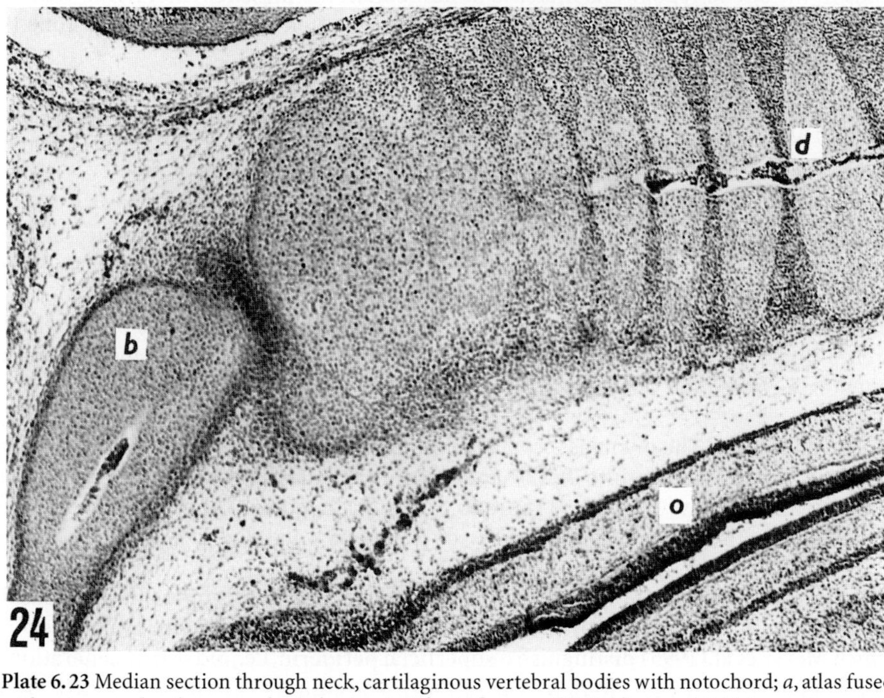

Plate 6. 23 Median section through neck, cartilaginous vertebral bodies with notochord; *a*, atlas fused with axis; notochord present also in basioccipitale. *P. phocoena* (MK 67), stage 6, TL 51 mm. Azan, ×40. **24** Median section through neck, near previous section, cranially. Notochord in *b*, basioccipitale; *d*, notochord expands between vertebral bodies; *o*, esophagus. *P. phocoena* (MK 67), stage 6, TL 51 mm. Azan, ×40

was shifted dorsorostrally resulting in a close apposition of both structures. Thus the long and wide fossa rhomboidea of earlier stages is now almost closed dorsally, thereby forming the short, definitive fourth ventricle. The primitive plexus choroideus is present in the membranous roof of the fourth ventricle. Hitherto cerebellar proliferation has been confined mainly to the intraventricular portion. Dorsally both primordial cerebellar hemispheres extend toward the mediosagittal plane but have not yet fused to form the vermis. Thus the dorsal alar plate remains thin.

Myelencephalon – The cervical flexure, i.e. the angle between brain base and spinal cord is slightly obtuse by now and it is an indication of progressive body straightening. Contrary to e.g. telencephalon, the myelencephalon has relatively shortened. At the same time the tegmentum of the caudal hindbrain is considerably thickened. Further caudally an extensive proliferation of the myelencephalic alar plate and spinal cord alar plate has narrowed down the aperture and lumen of the central canal. The ventral median fissure is slightly deepened by the expansion of the developing ventral columns.

Medulla Spinalis. The canalis centralis becomes gradually narrowed; in cross-section it is elongated in the dorsoventral direction and widened in its ventral part.

Hypophysis. The solid neurohypophysis has a fingerlike appearance, ventrally it is completely covered by the closely attached broad adenohypophysis. The epithelial folds in the adenohypophysis increase in number. In the mesenchymal tissue ventral of the hypophyseal primordium, the prochondral blastema of basisphenoid bone is differentiated.

Eye. The lens vesicle is completely filled with lens fibers. The epithelial buds of the conjunctiva protrude into connective tissue lateral of the eye as anlage of the lacrimal gland.

Skeletal System. Ossification centers and the first lamellae of bone tissue can be demonstrated in some of the desmal bones, such as the mandible, the maxilla, and the clavicle. Regions of hypertrophied cartilage are found in the cartilaginous models of ribs, vertebral arches, and bodies, and perichondrial ossification begins on vertebral arches and ribs. The dorsal cord is still continuous, uninterrupted, and becomes extended between the vertebral bodies.

Muscular System. The myoblasts continue increasing in length and the muscles show incipient differentiation into fleshy and tendinous parts. Mesodermal (muscular) abdominal walls have descended more to the ventral median plane, but between the walls there is still enough space for the wide umbilical stalk.

Integumentum Commune. The epidermis of the dorsum and flippers forms a basal layer of cuboid cells and one or two layers of polygonal to flat cells with oval nuclei, in which Meyer et al.(1995) distinguish a superficial periderm, i.e., older first generation, and a younger one, i.e., stratum intermedium-I. The authors have demonstrated such differentiation, albeit less distinct, even in the abdominal region. For the first time, a basal membrane is present in all body regions. Numerous fibroblasts are present in the dermis, and collagen fibrils and blood sinuses begin appearing (Meyer et al. 1995).

Lick (1987) distinguishes three layers of dermis in Pp: (1) under the epidermis there is a wider layer of loose connective tissue of embryonic character, with irregularly arranged fibroblasts. (2) Below it there is a thin layer of dense connective tissue in which the collagen fibrils and fibroblasts are arranged parallel with the surface. This layers passes into the panniculus carnosus (musculus cutaneus trunci), which is very well developed in Pp (Lick 1987) (3) Then follows a deeper thin layer with numerous fibroblasts and occasional collagen fibers. Lick (1987) has also demonstrated, in Pp (MK 62), epithelial buds of tactile hairs on the upper lip (two on the right and three on the left).

5.1.6.1 Material

- **Sa:** 24, 32, 35, 36, 37, 38, 39 (K 18), 40, 41, 42, 43, 44, 45, 46 (K 24), 47. TL=42–74 mm, CRL=19–32 mm, W=0.61–2.2 g.
- **Sl:** 11, 12, 13, 14, 15. TL=66–79 mm, CRL=27–35 mm, W=1.65–3.15 g.
- **Dd:** 8, 9, 10, 11. TL=52–69 mm, CRL=22–30 mm, W=0.82–1.7 g.
- **Pp:** 6 (MK 67), 7 (MK 64), 8 (MK 75), 9 (MK 62). TL=51–70 mm, CRL=22–29 mm.

5.1.7 Stage 7: Eyelids Fused, Umbilical Hernia Reposited

Estimated age, range 51–66 days, mean 57 days (Fig. 7, Plates 7, 8).

This stage, considered to be the first fetal one, is characterized by ontogenetic steps that isolate the fetus from its surroundings. They include the sealing of eyelids, external auditory meatus and nostrils, and the closure of the umbilical stalk following the reposition of the physiological umbilical hernia.

Externally. Fetuses show the typical dolphin appearance. The embryonic curvature is almost indistinct, the axes of the trunk and tail forming a considerably obtuse angle,

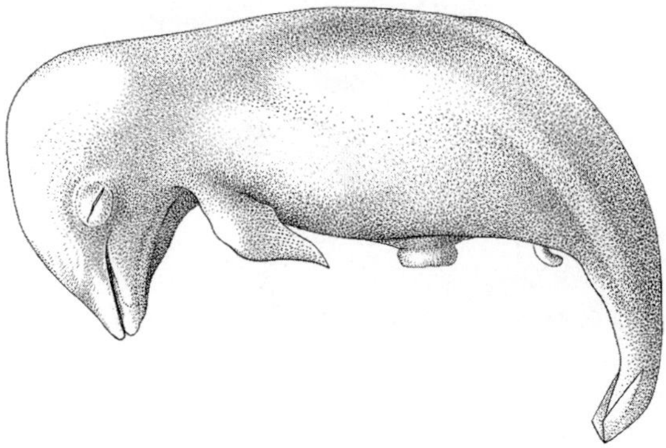

Fig. 7. External appearance of a *Stenella attenuata* fetus (*Sa 54*) in stage 7

25

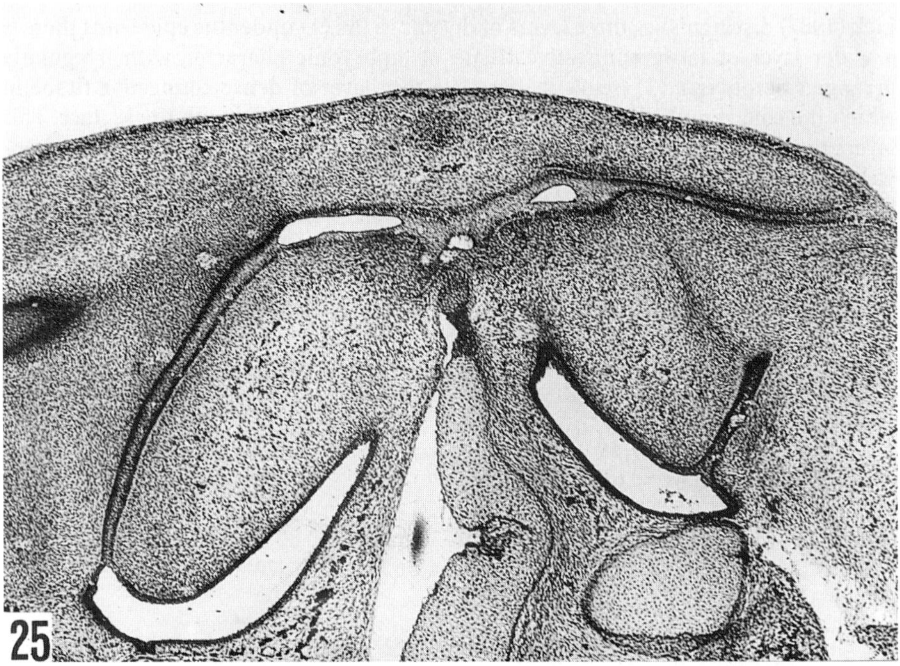

26

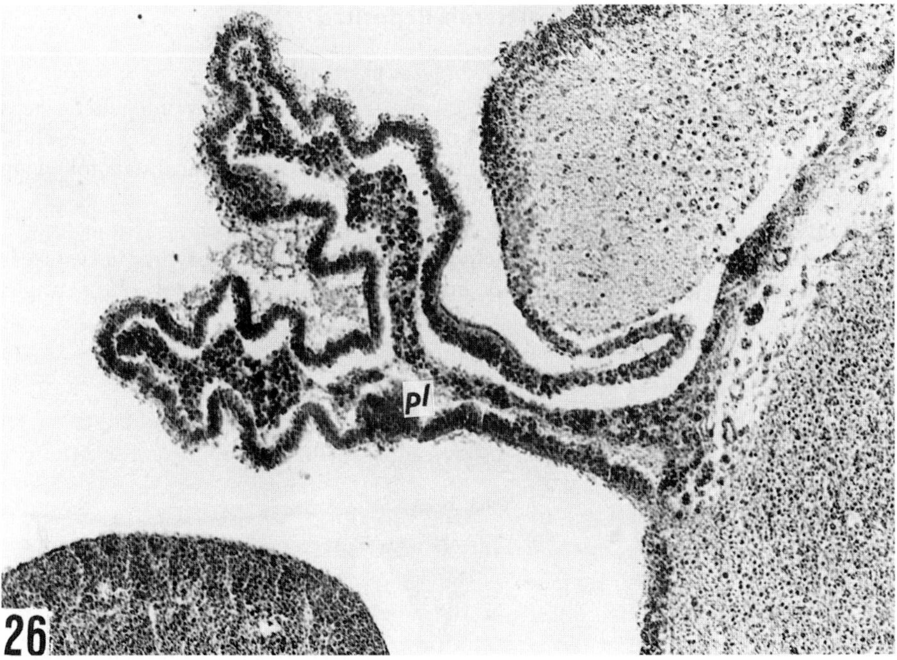

Plate 7. 25 Cross-section through asymmetric nostrils and meatus nasi. *P. phocoena* (MK 61), stage 7, TL 95 mm. Azan, ×100. **26** Cross-section of telencephalon; *pl*, plexus choroideus in lateral ventricle. *D. delphis* (K 49), stage 7, TL 82 mm. Azan, ×20.

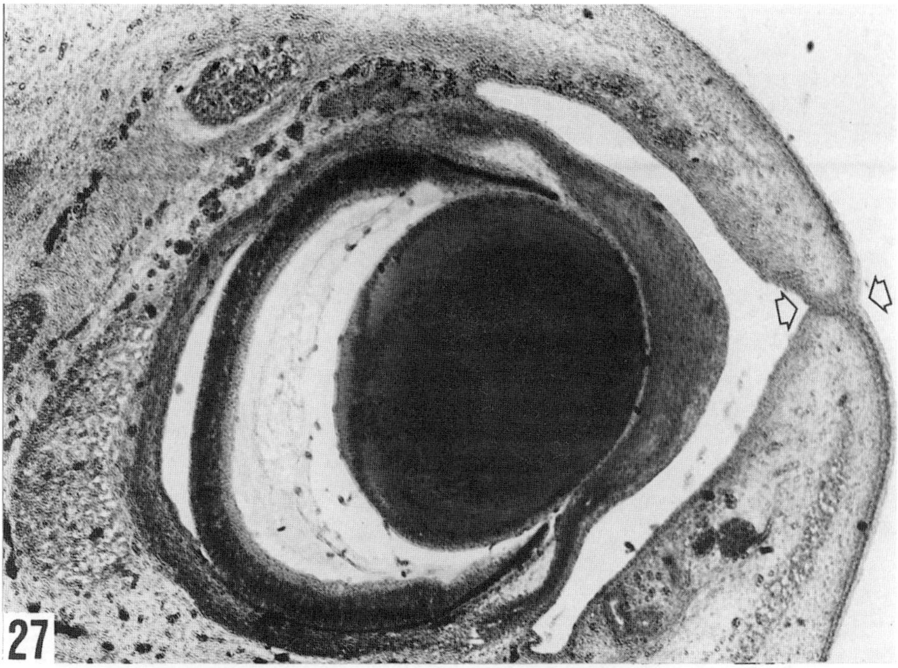

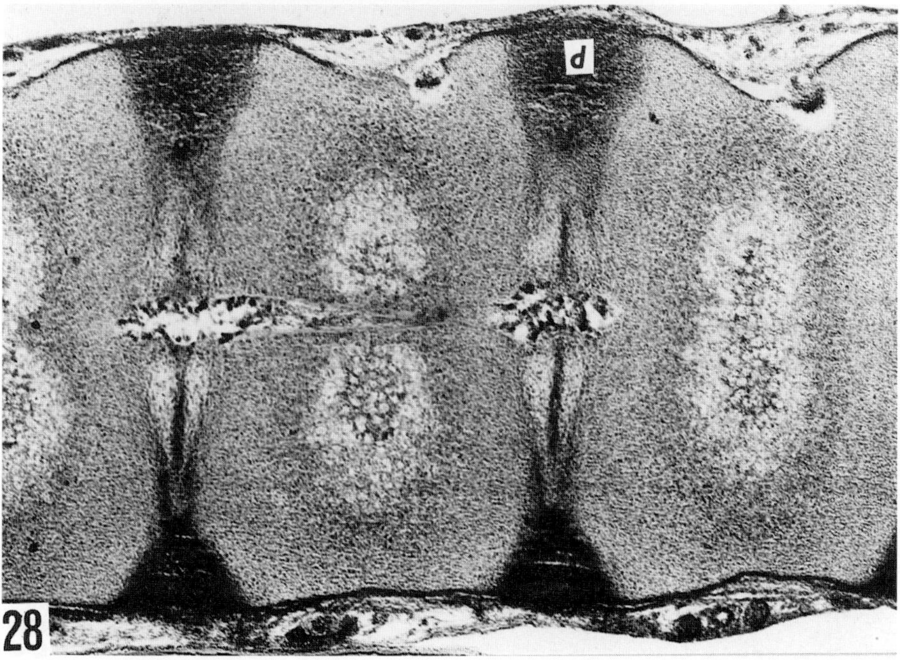

Plate 7. 27 Section through optic bulb, sealed lids (*arrows*). *P. phocoena* (K 19), stage 7, TL 90 mm. Azan, ×40. **28** Median section through thoracic vertebral bodies with hypertrophied and calcified cartilage; intervertebral discs are forming (anuli fibrosi, nuclei pulposi). *P. phocoena* (MK 61), stage 7, TL 95 mm. Azan, ×40

Plate 8. 29 Umbilical canal obliterated, intestinal loops in abdominal cavity; *u*, umbilical cord. *D. delphis* (K 49), stage 7, TL 82 mm. Azan, ×20. **30** Scapulohumeral joint, articular cavity (*arrows*); c, hypertrophied cartilage. *D. delphis* (K 49), stage 7, TL 82 mm. Azan, ×100.

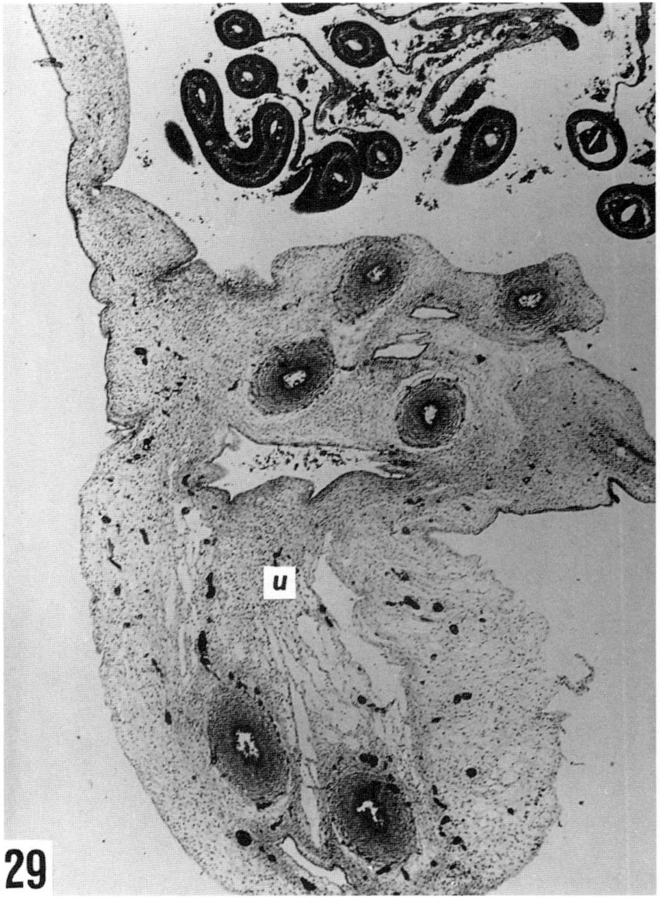

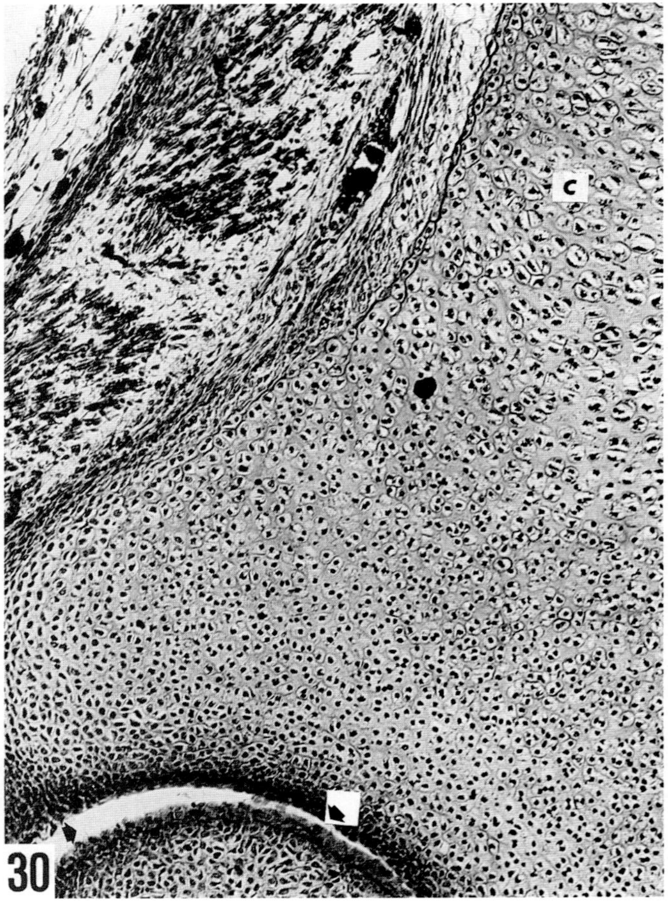

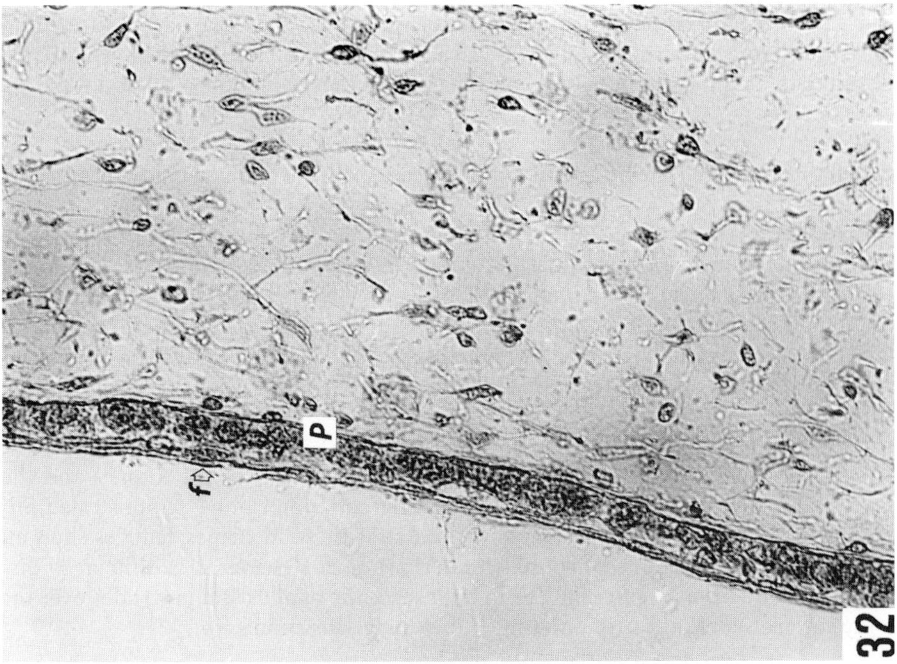

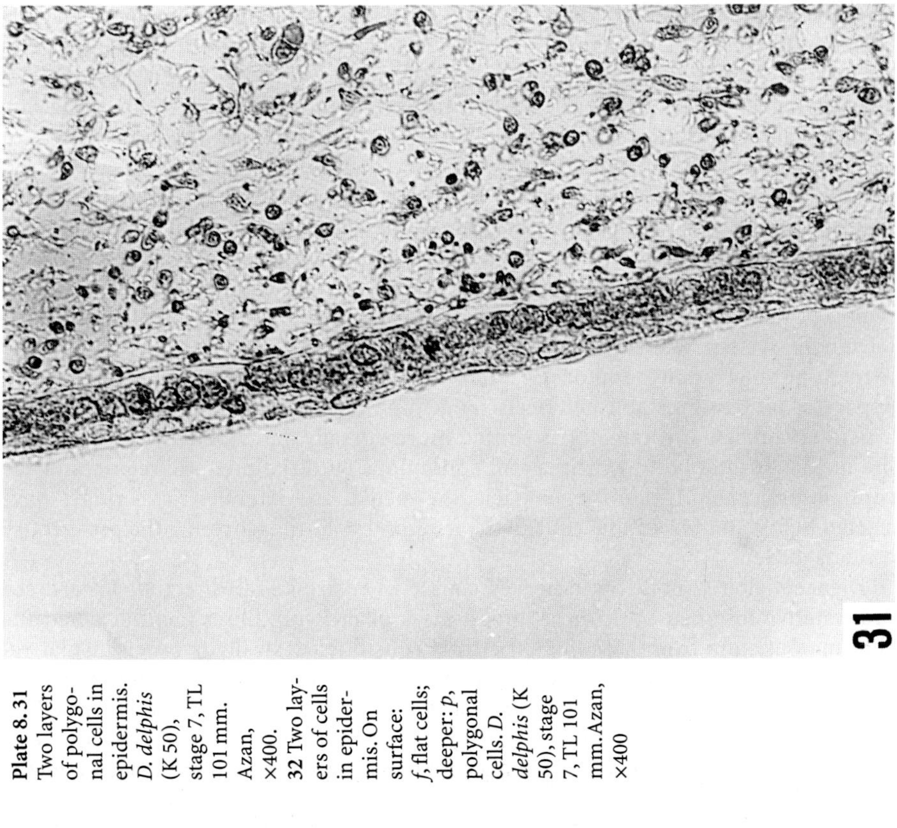

Plate 8.31 Two layers of polygonal cells in epidermis. *D. delphis* (K 50), stage 7, TL 101 mm. Azan, ×400.
32 Two layers of cells in epidermis. On surface: *f*, flat cells; deeper: *p*, polygonal cells. *D. delphis* (K 50), stage 7, TL 101 mm. Azan, ×400

those of the head and trunk an approximately right one. The facial part of the skull is long, the jaws delimiting a long and straight rima oris. The blowhole is situated dorsally on a connective line between the eyes. The growing dorsal fin and fluke are conspicuous.

Circulatory System. The heart and major vessels show the prenatal configuration.

Gastropulmonary System. On the dental lamina, the dental buds develop as a row of swellings. Due to the longitudinal growth of the fetus, the space in the abdominal cavity has been freed and the intestinal loops, previously situated in the wide umbilical stalk, have now been retracted into the abdominal cavity. The larynx contains laryngeal cartilage and assumes the form typical of Odontoceti. The cartilaginous rings are present in the tracheal wall and large bronchus. Bronchial ramification continues.

Urogenital System. Glomerule differentiation is under way in the kidney, and the ureter leads into the sinus urogenitalis. Sex differentiation can be demonstrated microscopically in the gonads. Males show characteristic solid stripes, females show an appreciably larger paramesonephric (müllerian) duct. However, it is difficult to distinguish between sexes externally, as in both sexes the phallus still protrudes over the body surface, its rounded tip pointing in the caudal direction.

Nervous System. Buhl and Oelschläger (1988) studied the morphogeny of the encephalon in a fetus of Pp at this stage, processed histologically in the series MK 69B. We quote their description:

Telencephalon – The external form of the brain is dominated by the towering telencephalic hemispheres. These have mainly expanded in four directions: rostrally, caudally, laterally and dorsally. In the lateral view they partly cover the rostral colliculi, whereas in the dorsal view the tectum remains uncovered. The cerebral hemispheres now quite obviously exceed the cerebellum in the width. The olfactory bulb anlage is obviously in a process of complete reduction. Remaining placodal material is gradually transformed to a ganglionic mass, with several rootlets attached to the septal region of the brain. The uncoupled ganglion is mainly composed of material of the terminalis system, which persists to adulthood. Therefore the anatomical arrangement typical of adult odontocetes, such as the absence of a peripheral olfactory apparatus is already established (Oelschläger and Buhl 1985; Oelschläger et al. 1987). The interventricular foramen has shifted more dorsally and is narrowed to a slit by the medial striatal body, which is still continuous with the preoptic region and the commissural plate. Inside the ventricle, the medial and lateral striatal ridges have started to fuse at their caudal ends in the temporal pole, thus forming the prospective caudate nucleus.

Diencephalon – Both optic nerves course in rostrolateral direction toward the presumptive foramen opticum. A broad mass of crossing fibres has thickened the chiasma opticum; from here the optic tract runs dorsorostrally between insular region and temporal pole. The dorsoventral extension of the hypothalamic floor has been remarkably reduced by a ventralward shift of the mamillary body caused by proliferating subthalamic brain centres. Concomitantly the height of the mesencephalic flexure is drastically reduced. Overall, the ventricular surface and the configuration of the diencephalic subunits have undergone some changes. Firstly the extension of the diencephalon in the rostrocaudal axis is reduced, whereas the dorsoventral extent has considerably increased. In consequence, epithalamus, thalamus and hypothalamus are arranged in a dorsoventral sequence. Secondly the thalamus, and predominantly the dorsal thalamus, now forms the dominant part of the diencephalic ventricular surface. Expansion of the dorsal thalamus has displaced the less progressive epithalamus to a more dorsal position. In the epithalamic roof, an inconspicuous pineal body anlage is present between the distinct habenular and epithalamic commissures. Most of the remaining epithalamic roof has been transformed to a thin chorioid plexus covering the third ventricle.

Mesencephalon – Externally the midbrain has lost its bulbous appearance. Two oblique grooves segregate the rostral from the caudal colliculi, the dominance of the caudal colliculi is not yet established. The caudal colliculi partly overlie the primitive cerebellar vermis and hemispheres, thereby markedly deepening and laterally expanding the former isthmus rhombencephali. Parallel to external differentiation, cellular proliferation has thickened the tectal walls, the width of the mesencephalic aqueduct is reduced. The most prominent feature is the remarkably developed tegmentum. The conspicuous cerebral peduncles can be recognized on both sides of the interpeduncular fossa (former mesencephalic flexure).

Metencephalon – Both cerebellar primordia have fused in the midline, giving rise to the vermis, thus thickening the formerly thin metencephalic roof. The latter process and the strong proliferation of rhombocephalic tegmental nuclei have further reduced the lumen of the fourth ventricle. In the dorsal aspect only a narrow, deep incision between cerebellum and myelencephalon indicates the presence of the fourth ventricle. Basally a marked rostral extension of the incipient pons, produced by proliferating pontine nuclei is recognizable.

Myelencephalon – In the course of body straightening the spinal cord emerges from the hindbrain almost parallel to the brain base. This is mainly attributable to two morphogenetic processes: first, a decrease in the pontine flexure angle accompanied by an increase in the cervical flexure angle. Second, at the cervical flexure, a conspicuous caudally oriented bulge is apparent. It is caused by the highly progressive medial accessory inferior olive, which protrudes even more in adult.

Medulla Spinalis. The narrowing of the central channel continues. In cross-section, the channel is somewhat elongated in the dorsoventral direction, embedded in the gray matter; one can discern both ventral commissures and also the fissura ventralis is distinct.

Hypophysis. The neurohypophyseal stalk has been shifted ventrally and rostrally toward the postoptic recess. The slightly flattened adenohypophysis has extended further laterally and caudally. A pars intermedia is missing. Histogenesis of both parts is in progress, the surrounding mesenchyma differentiates into structures of fossa hypophyseos.

Eye. The black-and-brown pigment in the iris is very conspicuous. The eyelids have grown over the eyeball, contacted and fused by an epithelial seal, the latter being still incomplete in the youngest specimens Sa 48 and Sa 49.

Skeletal System. The desmal bones of calva become ossified at this stage and also the vomer, maxilla, os incisivum, and mandible. The cartilaginous nasal skeleton still retains its cartilaginous character (Klima and van Bree 1985). The notochord shows extensions between vertebrae, and it disappears in the vertebral bodies in some places. In the axial skeleton, perichondrial ossification takes place in vertebral arches and ribs. On the diaphyses of long limb bones, either hypertrophied cartilage or thin cuffs of periosteal bone are found. Articular cavities appear in places of the prospective joints.

Muscular System. The myoblasts successively change into myotubes which assemble to form primary bundles. Also, the interstitial tissue begins to differentiate muscular septa, tendons, aponeuroses, and perimysium. The muscular tissue of the abdominal wall has grown up to the median plane, thereby narrowing the umbilical stalk which, up to this stage, has permitted the physiological hernia of intestinal loops.

Integumentum Commune. Fine, pale brown skin pigmentation begins to appear in Sa (Sa 49–Sa 57). The periderm has become very flat and the stratum intermedium-I shows two to five layers in the dorsal region, and four to five layers in the flippers. The stratum intermedium-I of the abdomen consists of one to three layers only. In all body regions, the basal layer starts the production of a third generation of epidermal cells: stratum intermedium-II (Meyer et al. 1995). Lick (1987) found, in the corium of Pp: (1) a thick layer of intertwined fibroblasts and collagen fibers; (2) a thin fascia that enters the cutaneous muscle, major blood vessels pass between these two layers; (3) a layer of loose tissue; (4) a thin layer of tough tissue; and (5) a thicker layer of loose tissue with numerous capillaries and nerve bundles. Lick (1987) found anlagen of tactile hairs in a fetus K 19 (three on the right and two on the left) and in MK 61 and MK 69 (two on the right and two on the left). The anlage includes a bulb and a hair in a hair sheath.

5.1.7.1
Material

- **Sa:** 48 (K 14), 49 (K 15), 50 (K 25), 51, 52, 53 (K 26), 54, 55, 56, 57, 58. TL=76–110 mm, CRL=35–55 mm, W=3.7–9.2 g.
- **Sl:** 16, 17, 18, 19. TL=88–97 mm, CRL=39–45 mm, W=4.9–6.7 g.
- **Dd:** 12 (K 49), 13, 14 (K 50). TL=82–114 mm, CRL=34–50 mm, W=3.3–10.6 g.
- **Pp:** 10 (K 19), 11 (MK 61), 12 (MK 69). TL=90–107 mm, CRL=36–46 mm.

5.1.8
Stage 8: Skin Folded

Estimated age, range 62–78 days, mean 70 days (Fig. 8; Plates 9, 10).

Externally. The body continues to straighten and even the angle between the head and the trunk changes from right to obtuse. The lower jaw (mandible) is usually longer than the maxilla. The fluke gradually enlarges laterally and caudally so that its caudal

Fig. 8. External appearance of a *Stenella attenuata* fetus (*Sa 61*) in stage 8

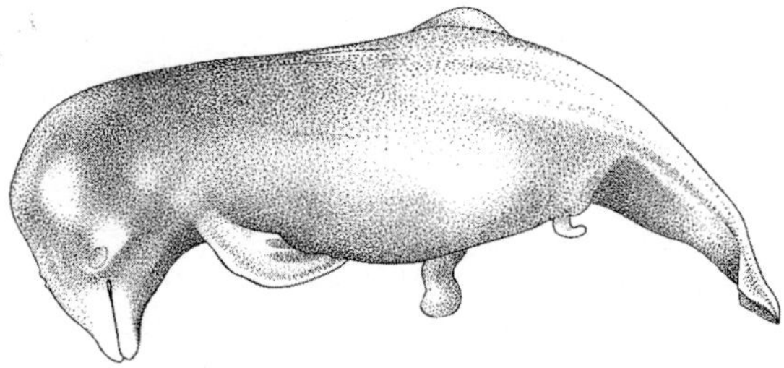

edge forms an arch with a caudal convexity. The anlage of the dorsal fin increases in height.

Circulatory System. No fundamental changes against the preceding stage are observed.

Gastropulmonary System. In the oral cavity, according to Míšek et al. (1996), the swellings of dental lamina are apparent also in the incisor area, and the dental buds increase in number (to approximately 24 in a jaw quadrant). Histogenetic changes take place in other organs.

Urogenital System. The kidneys show the lobular structure typical of cetaceans. In the renculi, the formation of corpuscles and tubules is still under way. Histogenetic differentiations continue in the gonads. Male gonads show a distinct thick tunica albuginea, septa, lobuli, and rete testis. Tubuli seminiferi contorti become differentiated. In females, the gonadal cords divide into numerous follicles, the tunica albuginea does not develop. Sex differences appear on the external genitals. The male phallus, the tip of which becomes pointed, is still turned caudally but is gradually covered with a prepuce, the opening of which is directed cranially. The female phallus is gradually covered with labia pudenda and its tip still points caudally.

Nervous System. A dorsal outgrowth of the telencephalon above the tectum indicates the progressive towering of the hemispheres. The olfactory bulb has been totally reduced. The cerebellar hemispheres have noticeably extended in the caudal direction, now partly overlapping the myelencephalon (Buhl and Oelschläger 1988). The neocortex is differentiated into the following layers starting from the surface: (1) molecular (plexiform) zone; (2) granular and outer pyramidal zone; (3) inner white zone; and (4) inner pyramidal and matrix zone.

Hypophysis. The glandular lobe is filled with folded epithelium, the primitive cavity of Rathke's pouch is negligible. The hypophysis lies in the fossa hypophyseos, the sphenoid bone still being cartilaginous. Some of the specimens still show vestiges of the adenohypophyseal stalk with epithelial lining.

Eye. In the eyeball, the development of inner structures and of accessory organs takes place.

33

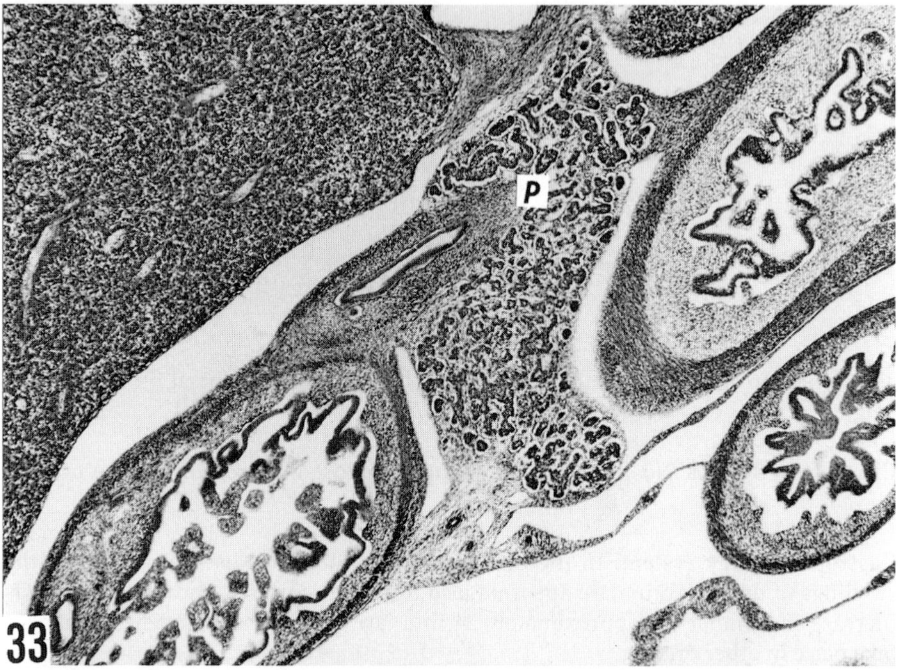

34

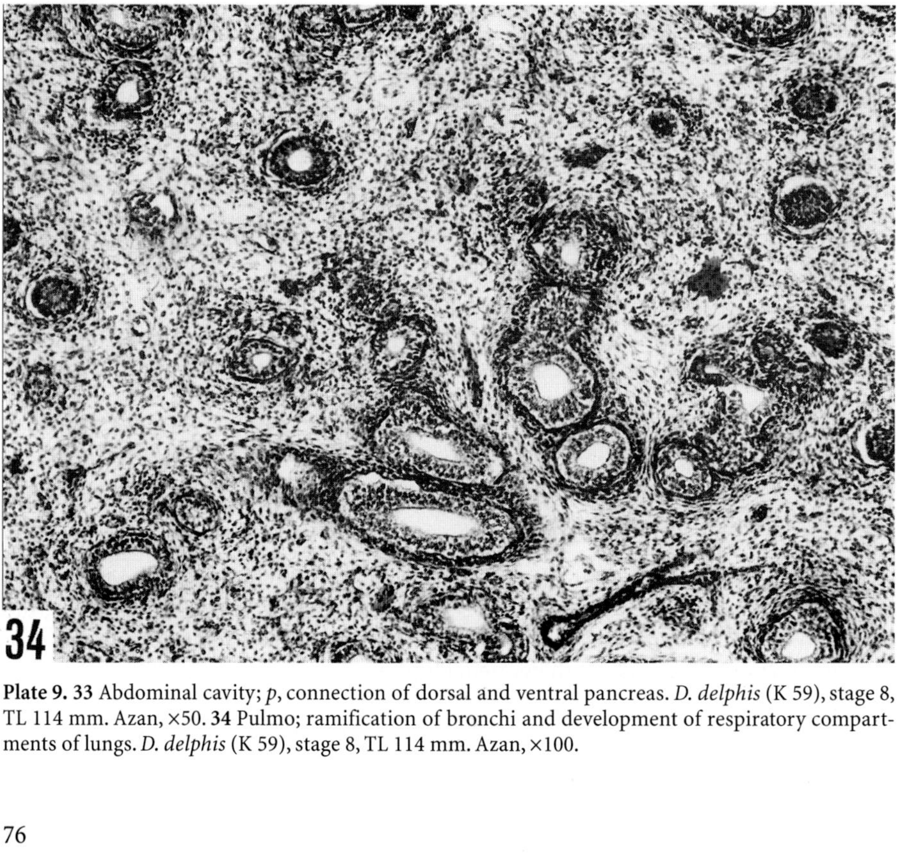

Plate 9. 33 Abdominal cavity; *p*, connection of dorsal and ventral pancreas. *D. delphis* (K 59), stage 8, TL 114 mm. Azan, ×50. **34** Pulmo; ramification of bronchi and development of respiratory compartments of lungs. *D. delphis* (K 59), stage 8, TL 114 mm. Azan, ×100.

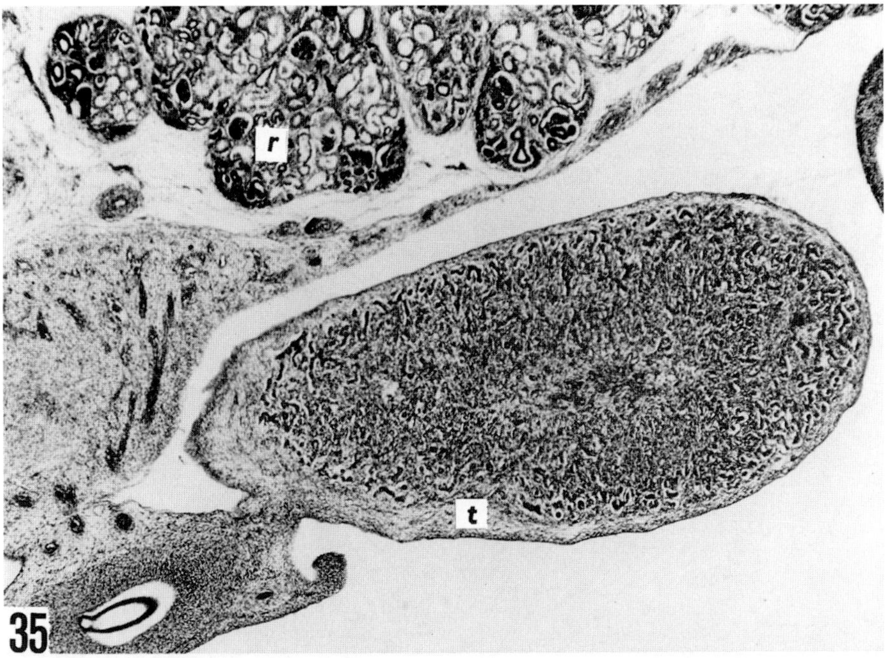

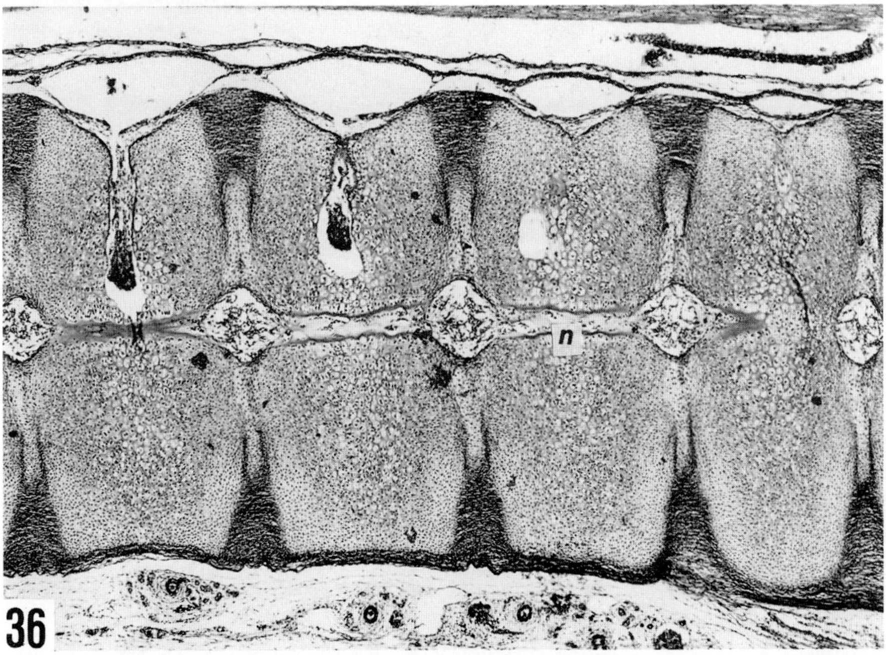

Plate 9. 35 Abdominal cavity; *r*, ren; *t*, testis, differentiation of seminiferous tubules and tunica albuginea. *D. delphis* (K 59), stage 8, TL 114 mm. Azan, ×50. **36** Median section through bodies of thoracic vertebrae, endosteal ossification begins, differentiation of intervertebral discs (nuclei pulposi and anuli fibrosi) progresses; *n*, remnants of notochord. *D. delphis* (K 59), stage 8, TL 114 mm. Azan, ×50

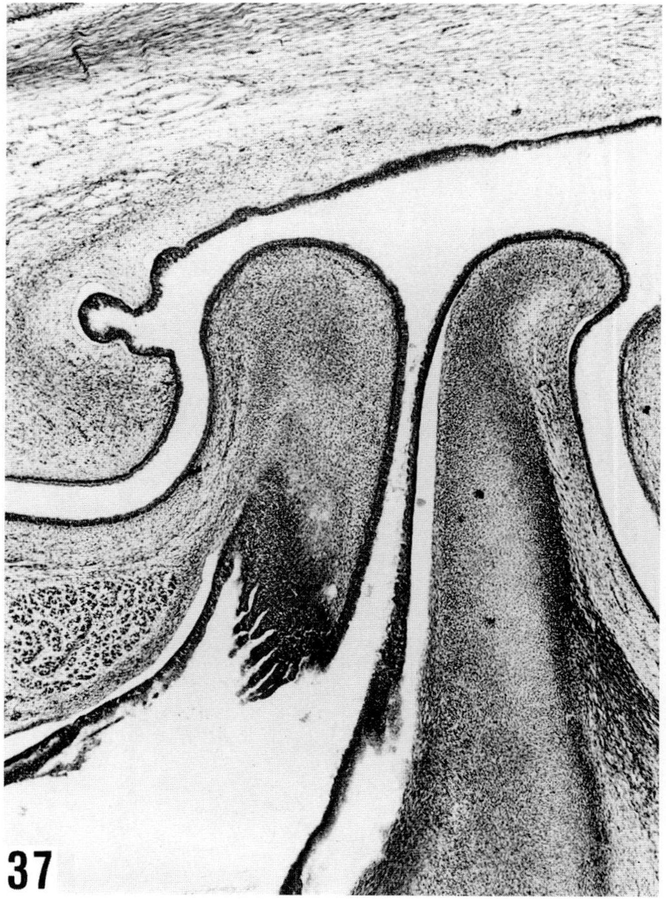

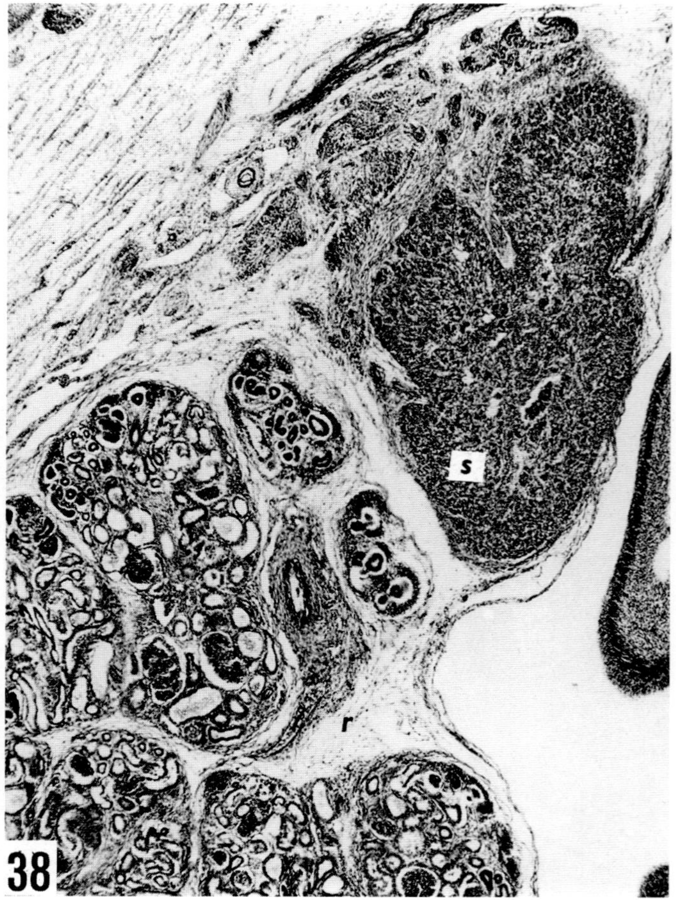

Plate 10.
37 Aditus laryngis in median section; its shape is already characteristic of dolphins. *D. delphis* (K59), stage 8, TL 114 mm. Azan, ×50.
38 Ren *(r)* and suprarenal gland *(s)*. *D. delphis* (K 59), stage 8, TL 114 mm. Azan, ×50.

Plate 10
39 Differentiation of neuronal layers in neocortex. *D. delphis* (K 59), stage 8, TL 114 mm. Azan, ×200.
40 Long myotubes in dorsal muscles. *D. delphis* (K 59), stage 8, TL 114 mm. Azan, ×200

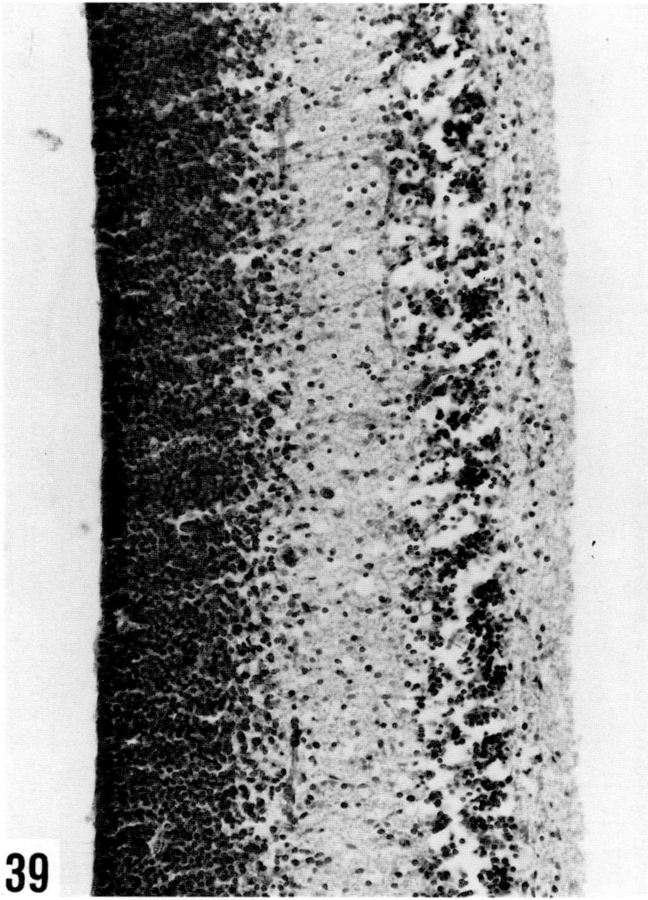

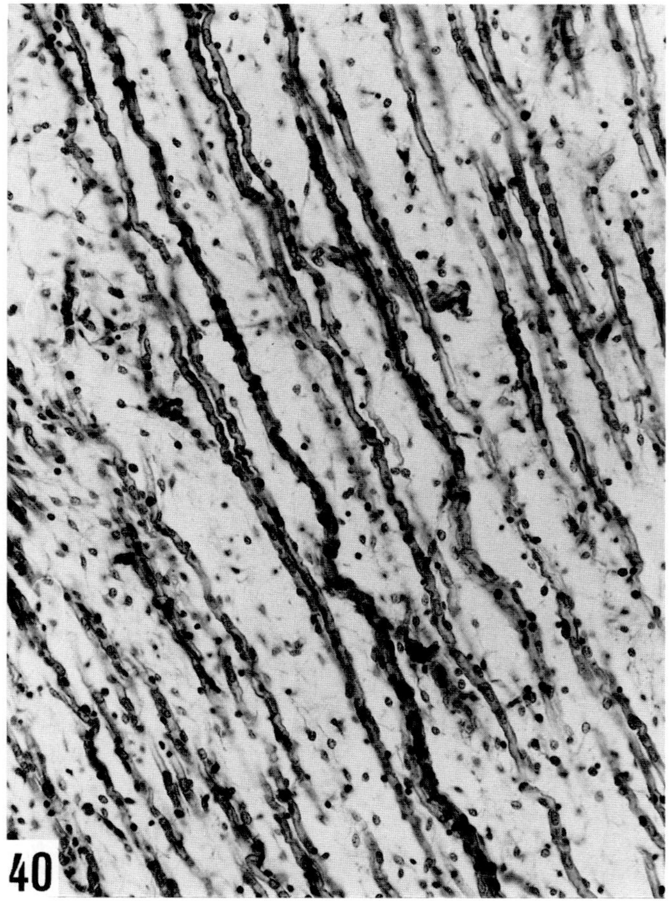

Skeletal System. The chordal canal is obliterated in some vertebrae. The spinal tips of neural processes join each other and form double spinal processes. The centers of vertebral bodies are built of hypertrophied cartilage cells. The periosteal ossification of long limb bones has only slightly advanced against the preceding stage; still no ossification has been demonstrated in the fetuses of Dd (K 45, K 50). The clavicle is subject to desmal ossification in Dd and Pp (Klima 1990).

Muscular System. The organogenesis of the muscular system proceeds at a very low rate, the changes against the preceding stage being hardly discernible.

Integumentum Commune. The number of epidermal layers remains the same as in the preceding stage or is only slightly increased. The skin increases in thickness, first of all, due to advancing growth and histogenesis of corium. The blood capillaries now approach the epidermis. The early subcutis (hypodermis) can be discerned by its wide-meshed lattice pattern (Meyer et al. 1995). In Pp (MK 76), Lick (1987) found, in the skin of each of the two upper lips, three anlagen of tactile hairs which did not differ in the degree of development from those found in fetuses at the preceding stage.

5.1.8.1
Material

- **Sa:** 59, 60, 61, 62, 63 (K 27), 64, 65, 66, 67. TL=112–134 mm, CRL=57–73 mm, W=12–21.4 g.
- **Sl:** 20, 21, 22, 23, 24. TL=114–137 mm, CRL=53–67 mm, W=10.7–23.9 g.
- **Dd:** 15 (K 59), 16, 17 (K 51), 18, 19. TL=114–144 mm, CRL=50–60 mm, W=10.6–22.3 g.
- **Pp:** 13 (MK 68), 14 (MK 76). TL=126–134 mm, CRL=55–60 mm.

5.1.9
Stage 9: Tactile Hairs Erupted

From this stage on, the estimated ages are given separately for the particular species and size categories in the paragraph on Material. Eruption of tactile hairs on the upper lip is the characteristic trait of this stage. The tactile hairs, being the only hairs in dolphins, erupt from the upper lip epidermis in Sa in the number of six to seven on either side, in Pp, two to three on either side (Fig. 9; Plates 11, 12).

Externally. The body shape is much the same as that of fetuses at the preceding stage, the lengths of jaws becoming even. The skin shows a slight brownish tint. The fluke changes in that its lateral lobes increase in size, its caudal edge changes from convex to convex–concave with an incision in the median line. The dorsal fin of older fetuses is triangular in shape. Phallus retraction continues in both sexes: in the males into the prepuce, as observed in fetuses Sa 70, Sa 71, Sa 75, Sa 76, and Sa 79; in females between the labia, as observed in fetuses Sa 68, Sa 69, Sa 72, Sa 73, Sa 74, Sa 77, and Sa 78.

Fig. 9. External appearance of a *Stenella attenuata* fetus (*Sa 72*) in stage 9

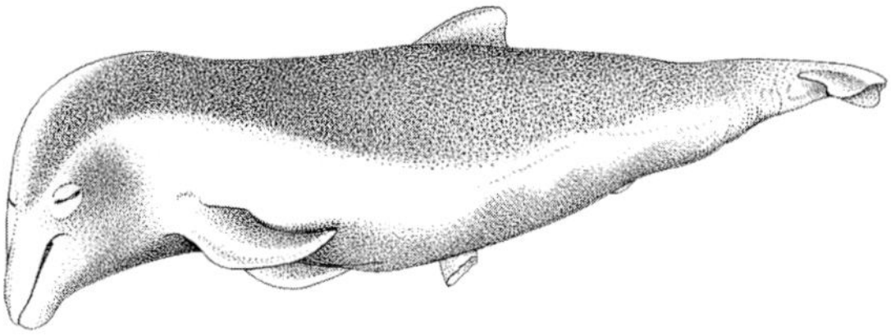

Gastropulmonary System. Some of the dental primordia reach the shape of dental cups, others remain in the shape of dental buds. Tonsillae develop in the form of numerous crypts with epithelial buds in the walls of the pharynx.

Nervous System. The differentiation of neocortical layers continues. Also the structures in falx cerebri (veins and venous sinuses) are forming.

Skeletal System. Neural processes are connected and form united spinal processes. In the axial skeleton, endosteal ossification of neural arches and ribs takes place. In vertebral bodies numerous vessels converge on their centers and enchondral ossification begins. Also, all parts of the chondrocranium are found in enchondral ossification. Desmal bones of the braincase and facial bones exhibit an advanced step of ossification.

Muscular System. The muscles are built of myotubes grouped in primary bundles.

Integumentum Commune. Meyer et al. (1995) report for Sa and Dd, that the dorsal and flipper epidermal intermediate-I layers are constantly of three to four in number. The abdominal stratum intermedium-I consists of five to six layers. The stratum intermedium-II shows three to four layers in the dorsum and abdomen and five to six layers in flippers. The intermediate-II cells are more slender and become distinctly flattened during the subsequent stages. The basal layer consists of high prismatic cells where the first melanin granules appear. The melanization continues in intermediate-II cells. The first small and rounded cells of the stratum spinosum (i.e., the fourth generation) are produced by basal cells. The dermis is characterized by an increase in blood vessel density. In Pp, according to Lick (1987), additional development takes place in the dermis: the author distinguishes five to six layers of corium. Beneath the epidermis there is a thick layer with dense collagen fibers in fine bundles, with major blood vessels and a distinct subepidermal capillary network. Beneath it lies a thicker layer of loose connective tissue, then a thin layer of tough connective tissue, the fascia, and finally the layer of loose connective tissue with a cutaneous muscle – musculus cutaneus trunci. Lick (1987) divides the fluke epidermis of Pp into three functional layers, viz., a stratum basale, a stratum spinosum, and a stratum corneum. The stratum basale consists of a layer of high prismatic cells, the lamina basalis is somewhat undulated but no corium papillae have developed as yet. Nor are any melanocytes or pigment granules present. The stratum spinosum consists of about eight layers, the stratum corneum of three or more layers of flat cells.

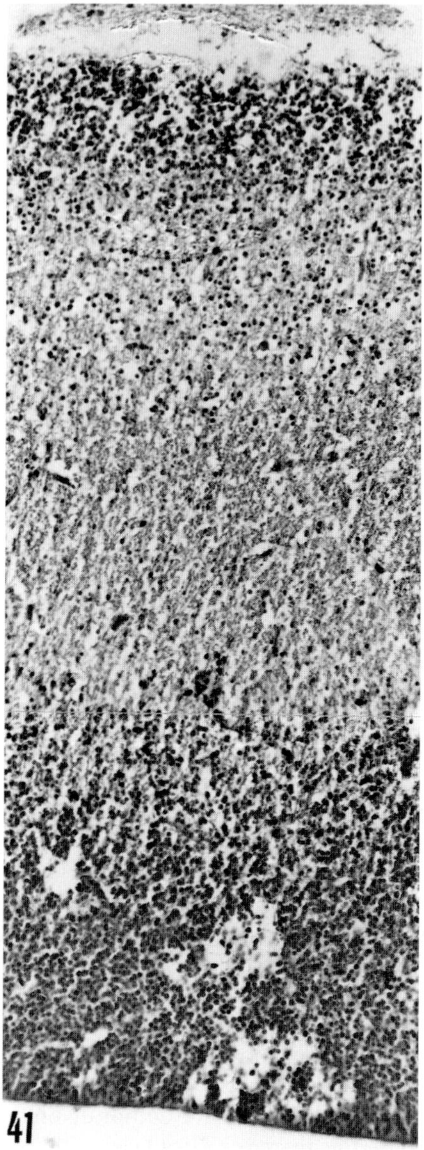

Plate 11. 41 Differentiation of neuronal layers in neocortex. *S. attenuata* (K 28), stage 9, TL 153 mm. Azan, ×200.

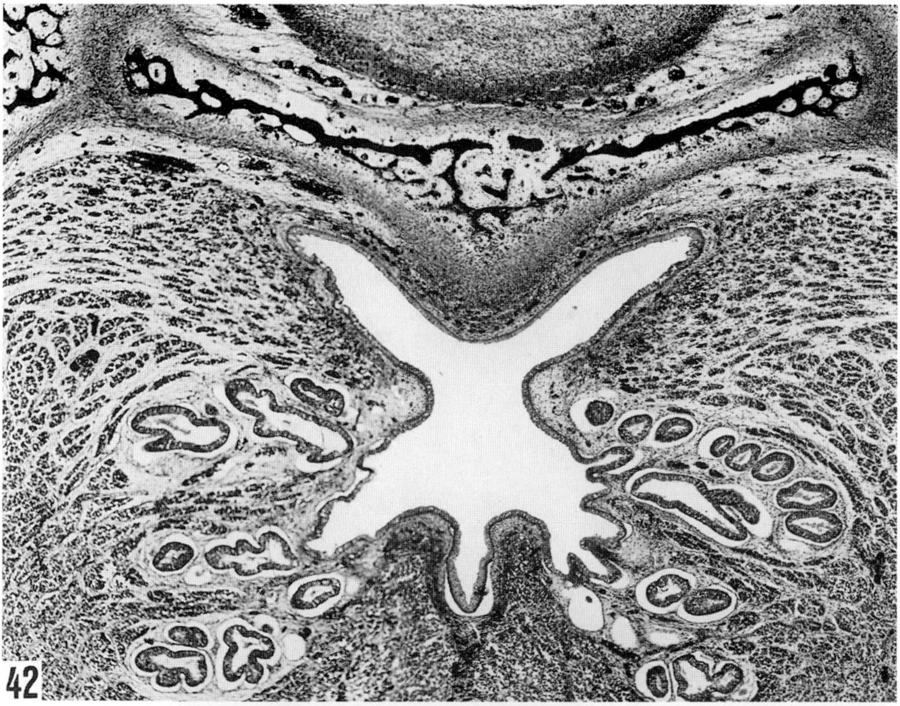

42

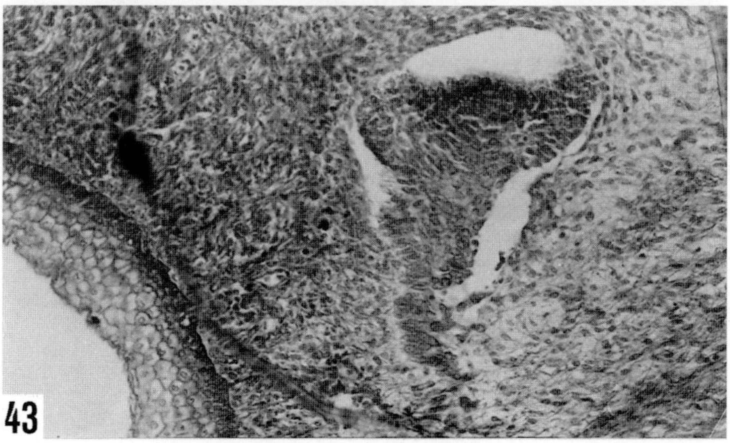

43

Plate 11. 42 Transversal section through meatus nasi. *S. attenuata* (K 35), stage 9, TL 225 mm. Azan, ×50. **43** Dental cup in upper jaw. *S. attenuata* (K 28), stage 9, TL 153 mm. Azan, ×200

Plate 12. 44 Transversal section through falx cerebri in *S. attenuata* (K 35), stage 9, TL 225 mm. Azan, ×50

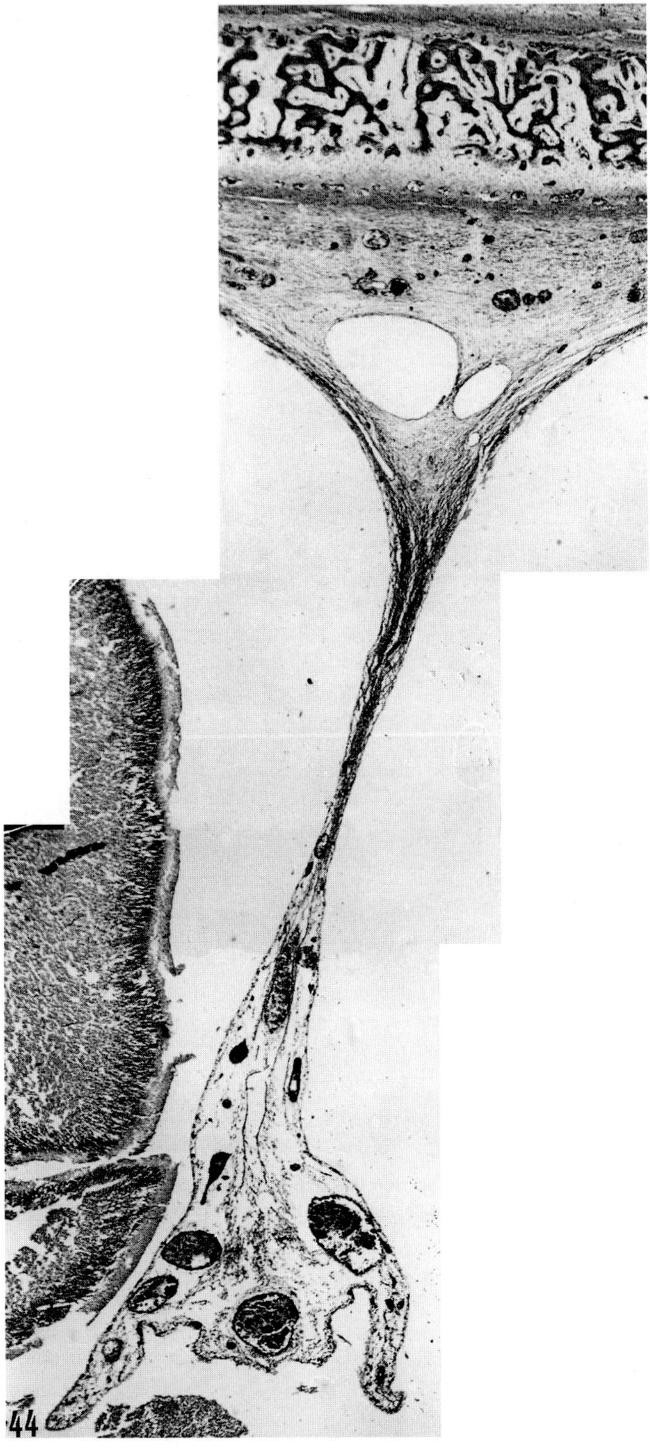

5.1.9.1
Material

Since the fetal growth is species-specific from this stage onward, and since the lengths (TL) and weights (W) of the fetuses show considerable variation ranges even within the same species, we have grouped them by 20–30 mm TL or 20–30 g W, and given the estimated age (EA) separately for each species and each size group.

- A. EA, range 78–85 days, mean 82 days.
  Sa: 68, TL=151 mm, CRL=80 mm, W=26.1 g; 69, TL=153 mm, CRL=82 mm, W=44.5 g, (K 28); 70, TL=157 mm, CRL=85 mm, W=36 g; 71, TL=165 mm, CRL=86 mm, W=36.5 g; 72, TL=167 mm, CRL=93 mm, W=44.5 g.
  Sl: 0
  Dd: 20, TL=163 mm, CRL=72 mm, W=32.8 g; 21, TL=164 mm, CRL=74 mm, W=36.8 g.
  Pp: 15, TL=167 mm, (MK 8, MK 9, MK 48, MK 49, MK 50, MK 53, MK 55, MK 59).
- B. EA, range 85–90 days, mean 87 days.
  Sa: 73, TL=187 mm, CRL=101 mm, W=64 g; 74, TL=192 mm, CRL=101 mm, W=54.1 g; 75, TL=193 mm, CRL=101 mm, W=52 g; 76, TL=205Nmm, CRL=108 mm, W=55.3 g.
  Sl: 25, TL=177 mm, CRL=93 mm, W=50 g, EA=85 days.
  Dd: 22, TL=194 mm, CRL=77 mm, W=55 g, EA=85 days.
- C. EA, range 90–100 days, mean 95 days.
  Sa: 77, TL=213 mm, CRL=102 mm, W=81 g; 78, TL=215 mm, CRL=110 mm, W=82 g; 79. TL=225 mm, CRL=112 mm, W=93 g, (K 35).
  Sl: 26. TL=198 mm, CRL=103 mm, W=90 g, EA=90 days.
  Dd: 0

5.1.10
Stage 12: Eyelids Open

At this stage, the epithelial seal of eyelids opens but the eyeball remains covered with eyelids for the most part. At the same time as the opening of the epithelial seal, a distinct brown coloration appears around the eyes and on the flipper. The tactile hairs on the upper lip show no appreciable growth, some of them remaining under the epithelial cover (Fig. 10).

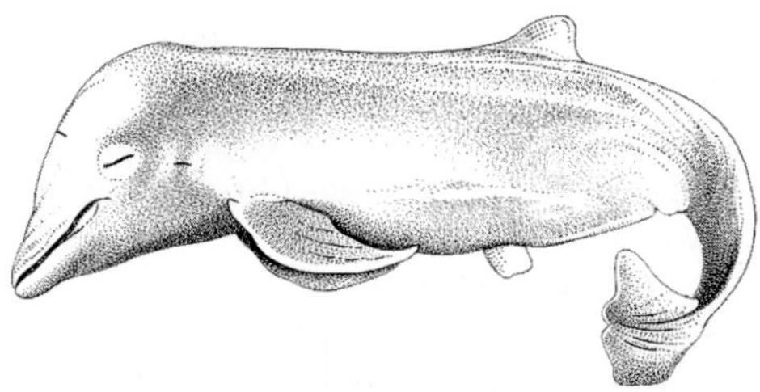

Fig. 10. External appearance of a *Stenella attenuata* fetus (*Sa 78*) in stage 12

5.1.10.1
Material

- **Sa:** 80, EA range=102–108 days, mean=104 days, TL=245 mm, CRL=132 mm, W=133 g.
- **Sl:** 27, EA range=95–105 days, mean=95 days, TL=215 mm, CRL=107 mm, W=89 g.
- **Dd:** 23, EA range=85–100 days, mean=90 days, TL=237 mm, CRL=116 mm, W=91 g.

5.1.11
Stage 10: First Body Hairs Erupt on the Body

As mentioned above, a haircoat which, at this stage, erupts from the epidermis in land mammals does not develop in cetaceans, nor are any anlagen found in them. In cetaceans this developmental stage is characterized by augmentation of pigment on the head and dorsum, imparting the fetuses with brown coloration, varying from pale brown in younger fetuses to deep brown in older ones. From this stage onwards, even the dorsal fins attain their species-specific shape (from Sa 83). In males, the retraction of glans penis into the prepuce is being finished (Sa 84, Sa 85); in females the clitoris is already covered by the labia (Sa 81, Sa 82, Sa 83) (Fig. 11).

Integumentum Commune. The periderm and stratum intermedium-I vanish and the number of layers increases in the stratum spinosum. Epidermal ridges develop (Meyer et al.1995). At this stage, the weights of the fetuses vary from 100 to 1000 g.

5.1.11.1
Material

- **A.** TL=25l–300 mm
  Sa: 81, EA range=105–110 days, mean=105 days, TL=260 mm, CRL=143 mm, W=138 g; 82, EA range=110–125 days, mean=115 days, TL=295 mm, CRL=152 mm, W=178 g.
  Sl: 28, EA range=100–130 days, mean=100 days, TL=244 mm, CRL=108 mm, W=113 g.
  Dd: 24, EA range=110–120 days, mean=115 days, TL=298 mm, CRL=128 mm, W=231 g; 25, EA range=135–145 days, mean=140 days, TL=300 mm, CRL=165 mm, W=456 g.
- **B.** TL=301–400 mm
  Sa: 83, EA range=135–150 days, mean=140 days, TL=394 mm, CRL=200 mm, W=397 g.
  Sl: 29, EA range=130–160 days, mean=135 days, TL=320 mm, CRL=137 mm.
  Dd: 26, EA range=130–140 days, mean 135 days, TL=320 mm, CRL=172 mm, W=394 g.
- **C.** TL=401–500 mm
  Sa: 84, EA range=160–175 days, mean=170 days, TL=468 mm, CRL=220 mm, W=693g; 85, EA range=160–175 days, mean=170 days, TL=477 mm, CRL=230 mm,

Fig. 11. External appearance of a *Stenella attenuata* fetus (*Sa 84*) in stage 10

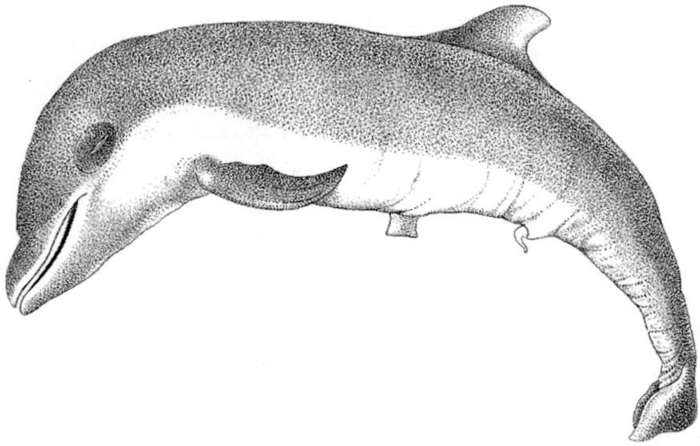

W=719 g.
Sl: 0
Dd: 27, EA range=160–175 days, mean=165 days, TL=482 mm, CRL=210 mm, W=854 g.

5.1.12 Stage 11: Haircoat all Over the Body

In dolphins, the haircoat is replaced by black to blackish-blue skin pigmentation, this coloration being apparent from the very beginning of this stage. Fetal Dd show the typical pattern resembling a horizontal figure "8" and a dark area around the eye (Fig. 12).

Integumentum Commune. The stratum intermedium-II has been lost completely so that the fetal epidermis exhibits a definite adult-like structure with supranuclear melanin caps in all cell layers. Distinct and elongate epidermal ridges project deeply into dermis. The stratum spinosum comprises eight to nine layers in the dorsum, or ten to 15 layers in flippers, which are covered by one or two layers of the first flat cells of stratum corneum. The dermis shows an increasing collagen fiber production with a generally denser structuring. In the adult-like dermis system, the collagen fiber bundles appear very massive in the stratum reticulare, as directly connected with the fiber bundles of the hypodermis. The hypodermal thickness develops more rapidly than the dermal thickness; the hypodermis occupies about 80–90% of the total integument (Meyer et al. 1995). The weights of the fetuses vary from 1010 g to over 6950 g.

5.1.12.1 Material

- **A.** TL=501–600 mm
  Sa: 0.
  Sl: 30, EA range=190–220 days, mean=210 days, TL=590 mm, CRL=230 mm.
  Dd: 28, EA range=175–195 days, mean=185 days, TL=580 mm, CRL=240 mm, W=1250 g.

Fig. 12. External appearance of a *Stenella attenuata* fetus (*Sa 87*) in stage 11

- **B.** TL=601–700 mm
  Sa: 86, EA range=190–210 days, mean=195 days, TL=615 mm, CRL=300 mm, W=1525 g; 87, EA range=200–220 days, mean=205 days, TL=660 mm, CRL=340 mm, W=1945 g; 88, EA range=210–230 days, mean=215 days, TL=690 mm, CRL=330 mm, W=2105 g.
  Sl: 31, EA range=220–245 days, mean=245 days, TL=700 mm, CRL=350 mm.
  Dd: 0
- **C.** TL=701–800 mm
  Sa: 89, EA range=220–240 days, mean=235 days,TL=795 mm, CRL=400 mm, W=3225 g.
- **Sl:** 32, EA range=245–270 days, mean=250 days, TL=720 mm, CRL=320 mm.
  Dd: 0
- **D.** TL=801–900 mm
- **Sa:** 0
  Sl: 33, EA range=270–290 days, mean=280 days, TL=805 mm, CRL=380 mm.
  Dd: 29, EA range=250–270 days, mean=255 days, TL=830 mm, CRL=420 mm, W=4010 g.
- **E.** TL=901–1000 mm
  Sa: 90, EA range=260–290 days, mean 270 days, TL=945 mm, CRL=510 mm, W=6950 g.

5.2 Development of Body Shape and Fetal Growth

Embryos in stages S3–S5 are curved, which is typical of mammalian embryos. They become gradually straightened from stage S6 on, and their body gradually assumes the elongated, caudally tapering shape typical of the family Delphinidae. We have followed up the growth of individual body parts and increase in weight in relation to total length (proportional or allometric growth). The growth of total length and individual body parts was also followed in relation to time, i.e., to the duration of intrauterine development (allochronic growth) and allometric growth in time. The calculations have been made only in *Stenella attenuata*, of which species the most numerous material was available to us. The dimensions are given in Tables 1, 5, 6, and 12; the numbers of dimensions studied are given in parentheses as follows: crown–rump length, CRL (1); total length, TL (2); crown–genitals (3); genitals–fluke (12); girth (21), and weight, W (22).

Table 12. Selected somatic dimensions of *Stenella attenuata* (1, 2, 3, 12, 21, 22)

Stage	Range	Σ	*n*	x	SD	Sx
1. Crown–Rump length						
3	8–10	36	4	9.0	0.7071	0.3535
4	11–13	94	8	11.7	0.8291	0.2931
5	14–21	313	18	17.4	1.9189	0.4523
6	19–32	393	15	26.2	3.9699	1.0250
7	35–55	500	11	45.5	5.6466	1.7025
8	50–73	540	9	60.0	6.6165	2.2055
9a	80–93	426	5	85.2	4.4452	1.9879
9b	101–108	411	4	102.7	3.0311	1.5155
9c	102–112	324	3	108.0	4.3205	2.4945
2. Total length						
3	21–25	89	4	22.2	1.6394	0.8197
4	28–39	243	7	34.7	3.6922	1.3955
5	33–45	702	17	41.3	3.1204	0.7568
6	42–74	893	15	59.5	10.4170	3.8729
7	76–110	1062	11	96.5	10.4217	3.3166
8	99–135	1070	9	118.9	8.1301	2.7100
9a	151–167	793	5	158.6	6.3749	2.8509
9b	187–205	777	4	194.2	6.6095	3.3048
9c	213–225	653	3	217.7	5.2493	3.0308
3. Crown–Genitals						
3	6.0–7.4	27.4	4	6.85	0.6191	0.2692
4	8.2–11.0	76.5	8	9.56	1.0914	0.3833
5	11.8–17.3	245.2	17	14.42	1.5566	0.5987
6	15.7–29.2	311.5	14	22.25	3.9627	1.5241
7	41.3–51	450.7	10	45.07	3.4948	1.8208
8	51.1–70.6	503.1	9	55.9	7.3612	3.1732
9a	77–110	406	5	81.2	5.1672	2.2722
9b	102–110	409	4 102.3	5.4392	2.1703	
9c	101–111	320	3 106.7	5.1316	2.9006	
12. Genitals–Fluke						
3	3.2–5.8	24	5	4.8	1.0296	0.3677
4	5.2–6.2	39.2	7	5.6	0.4000	0.1481
5	5.1–7.1	107.9	18	5.99	0.7149	0.2455

Table 12. *continued*

Stage	Range	Σ	*n*	x	SD	Sx
6	5.3–12.6	122.5	14	8.75	2.5282	0.9364
7	18.3–30	229.8	10	22.98	3.8064	1.7510
8	23–35.4	260.8	9	28.98	4.1242	2.0308
9a	37–45	204	5	40.8	3.0331	1.1665
9b	46–58	202	4	50.5	5.4467	2.3338
9c	49–64	164	3	54.67	8.1445	2.8085
21. Girth						
3	9–10	46	5	9.2	0.4000	0.1780
4	10–12	87	8	10.9	0.7806	0.2759
5	13–18	285	18	15.8	1.5000	0.3535
6	19–28	314	14	22.4	2.6380	0.7050
7	37–45	391	10	39.1	3.9610	1.2526
8	45–67	502	9	55.8	6.8114	2.2704
9a	66–88	384	5	76.8	6.9685	3.1163
9b	85–98	360	4	90	4.8477	2.4238
9c	105 –108	320	3	106.7	1.2472	0.7201
22. Weight						
3	0.035–0.09	0.323	5	0.0646	0.0206	0.0092
4	0.079–0.22	1.112	8	0.139	0.0558	0.0197
5	0.19 –0.62	8.05	18	0.4472	0.1200	0.0283
6	0.61 –2.2	18	14	1.2857	0.4580	0.1224
7	3.7–9.2	67.3	10	6.73	2.0377	0.6947
8	7.3–21.4	136.3	9	15.1	3.3483	1.1161
9a	26.1–44.5	187.6	5	37.5	6.8001	3.0410
9b	52.2–64	225.6	4	56.4	4.5249	2.2625
9c	81–93	256	3	85.3	5.4365	3.1388

x, mean; SD, standard deviation; Sx, standard error of the mean.

Fig. 13. Regression lines of allometric growth of body proportions in *S. attenuata*; logarithmic plotting. Ordinate: 1, crown–rump length (CRL); 3, crown–genital distance; 21, girth (circumference of thorax). Abscissa: 2, total length (TL; mm)

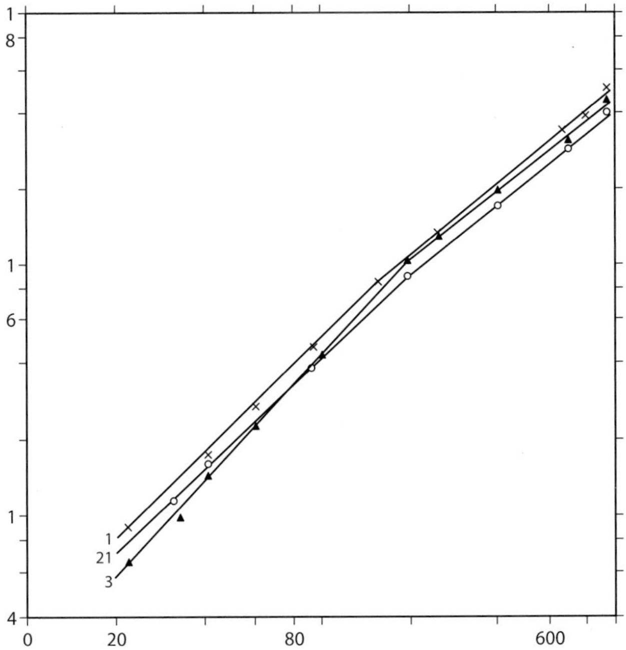

Fig. 14. Regression line of allometric growth of body proportions in *S. attenuata*; logarithmic plotting. Ordinate: 12, genitals-fluke; abscissa: 2, total length (TL; mm)

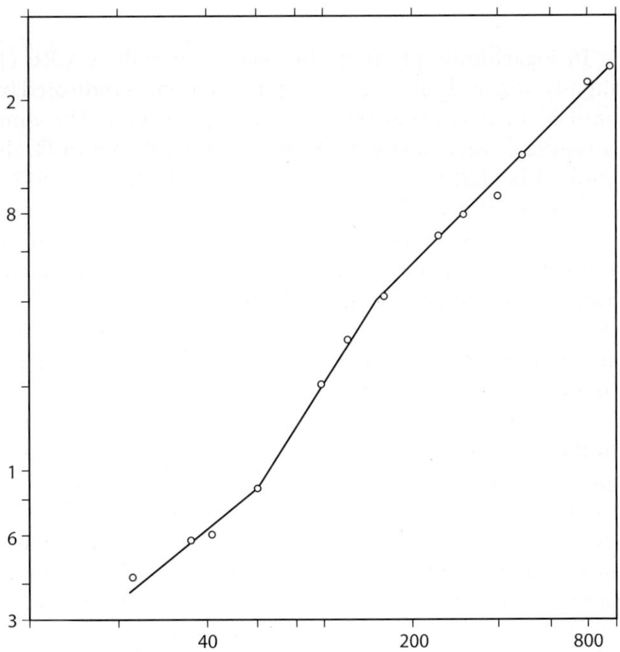

Fig. 15. Regression curve of allometric growth of weight (W) in *S. attenuata*; arithmetic plotting. Ordinate: 22, W (kg); abscissa: 2, total length (cm)

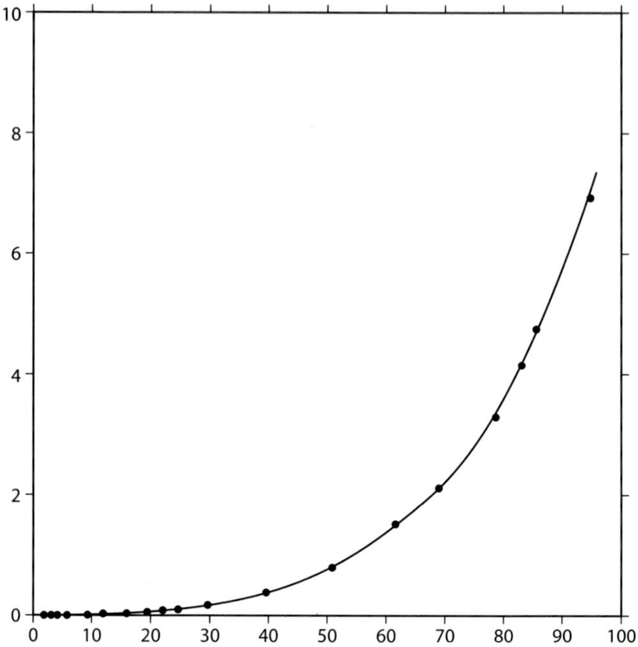

In logarithmic plotting, the relative growth of CRL (1) in relation to TL (2) is slightly positively allometrical up to stage S9, as indicated by the slope a=1.2; from S12 until birth it is negatively allometrical (a=0.9). The dimension of crown–genitals (3),which is very close to CRL, increases in relation to TL also positively allometrically until S9 (a=1.25) and negatively allometrically from S12 on (a=0.97) (Fig. 13). The dimension of genitals–fluke (12) in relation to TL increases negatively allometrically during the embryonic period up to S6 (a=0.8). From S6 to S9 it increases positively allometrically (a=1.7), and from S9 on isometrically (a=1.0) (Fig. 14). According to Schildger (1986), the girth (21) shows nearly isometrical growth in relation to TL (S3–S9 a=1.15, S9–S11 a=0.9). Weight (22), related to TL, shows a strongly positive allometrical growth throughout the whole intrauterine period (Fig. 15), slope a=3.4 (in logarithmic plotting).

In relation to time, the growth shows different relations in some cases (allochrony). In the arithmetic coordinate system, the growth curve of TL (2) shows a continuous increase (Fig. 16). In logarithmic plotting (Fig. 17), the TL (2) of the embryo increases, from 24 to 95 days of estimated age (EA), i.e., in S3–S9, positively allochronically (slope a=1.8). From 96 to 280 days of EA, i.e., from S9 until birth, the growth rate is slowed down a little (a=1.4). The CRL (1) shows a positive allochronical growth (a=2.20) from 24 to 95 days EA, i.e., in S3–S9. From 96 to 280 days, i.e., from S12 until birth, the growth slows down (slope a=1.3). The very similar dimension, crown–genitals (3), shows a resembling positive allochronical growth in S3–S9 (a=2.2), and from S12 until birth also slows down (slope a=1.3, Fig. 17). The genitals–fluke dimension (12) shows a negatively allochronical growth (a=0.5) from 24 to 38 days EA, i.e., in

Fig. 16. Regression curve of allochronic growth of total length (TL) in *S. attenuata*; arithmetic plotting.
Ordinate: TL (cm); abscissa: time (days) of prenatal period

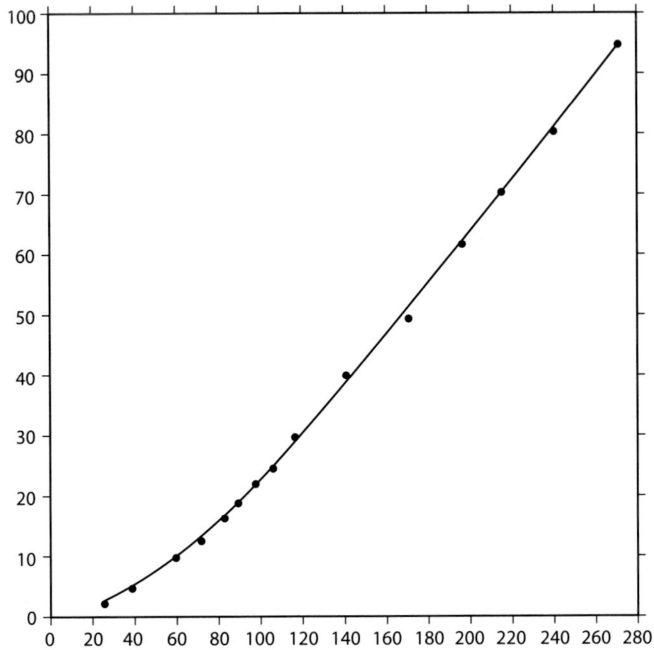

Fig. 17. Regression lines of allochronic growth of body dimensions (proportions) in *S. attenuata*; logarithmic plotting. Ordinate: 1, crown–rump length; 2, total length; 3, crown –genital distance; abscissa: time (days)

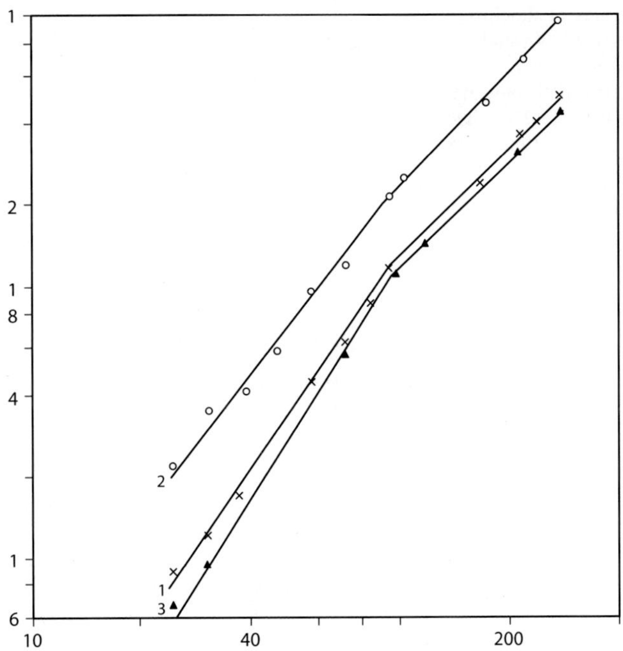

Fig. 18. Regression lines of allochronic growth of body proportions of *S. attenuata*; logarithmic plotting. Ordinate: 12, genitals–fluke; 21, girth; abscissa: time (days)

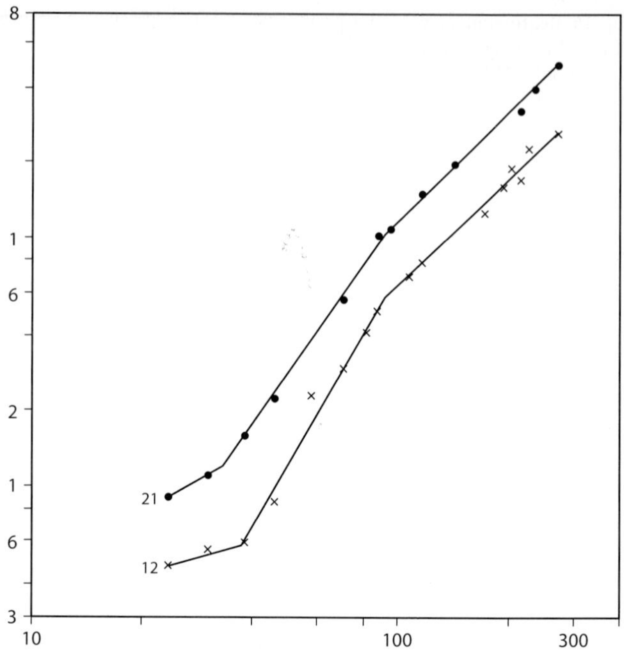

Fig. 19. Regression curve of allochronic growth of weight of *S. attenuata*; arithmetic plotting. Ordinate: 22, weight (kg); abscissa: time (days)

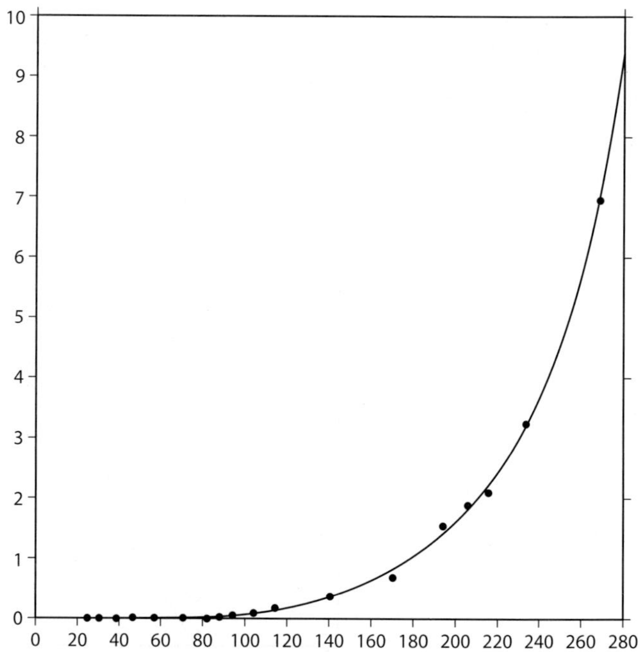

S3–S5, and a positive allochronical growth (a=2.6) in 39–95 days EA, i.e., in S5–S9. Then, between 96 and 280 days EA, i.e., in S9–S11, the growth rate drops to a=1.6. (Fig. 18). The girth dimension (21) related to time also shows a negative allochronical growth from 24 to 30 days (a=0.8) and a gradual positive allochronical growth (a=2.1) from 31 to 95 days EA, i.e., from S4 until S9, thereupon, the slope of its growth becomes more gentle, a=1.4 (Fig. 18). The growth curve of weight (22) related to time shows an intense increase (Fig. 19).

Also the allometric growth of the weight (W/TL) related to duration of intrauterine life (t), increases rapidly from the beginning of the fetal period on (Štěrba et al. 1994).

In logarithmic plotting, the weight shows a strong positively allochronical growth, from 24 to 95 days EA (S3–S9) a=5.5, from 95 to birth (S12–S11) it slows slightly to a=4.2.

5.3 Development of Head Shape and its Growth

At S3, the shape of the cephalic end of the embryo is spherical, rostrally showing lower paired tubercles of telencephalon and a high, unpaired tubercle of mesencephalon. Branchial bars are still visible in the developing facial region. In S4, the development of the facial region continues in gradual approximation and fusion of the nasal and maxillary processes; in S5 the nasolacrimal furrow is closed and the main facial features, such as the nasal pits, the eyeballs, the porus acusticus externus, and the mouth, are already distinct. The nostrils approximate mutually and are 2 mm distant from the rostrum. In S6, the palatine processes elevate and fuse, thus constituting the secondary palate. The eyelid anlagen appear. From S7 on, the facial part becomes conspicuously elongated, the head assuming the shape typical of the family and species. We have followed-up the dimensions on the head (Tables 5, 13) – rostrum–nostrils (5), rostrum–ear (6), rostrum–eye (8), eye–ear (10), nostrils–eye (11) – in relation to TL (2) and time, as well as the mutual relations between some of the dimensions (Schildger 1986; Štěrba et al. 1994).

The rostrum–ear distance (6), representing the skull length to a certain extent, shows positively allometrical growth in relation to TL, the growth rate decreasing gradually: S3–S8, a=1.76; S8–S11, a=1.09. The rostrum–eye distance (8), which may be considered to be the length of splanchnocranium, shows a positive allometric growth when related to TL, with a decreasing tendency, as indicated by the values of the slope: S3–S7, a=1.88; S8–S11, a=1.14. The eye–ear distance (10), which may be considered to be an index of neurocranium length, related to TL, shows slightly negative allometrical growth: S3–S12, a=0.94; S12–S11, a=0.84 (Fig. 20). The rostrum–nostrils distance (5), indicating the shifting of nostrils from rostrum to the crown, shows positively allometric growth to TL, with a decreasing tendency: from S3 to S7, a=2.28; from S8 to S11, a=1.11. The nostrils–eye distance (11), related to TL, also shows a decreasing tendency from allometrically positive to negative: S3–S7, a=1.8; S7–S11, a=0.9 (Fig. 21).

Related to time, the rostrum–ear dimension (6) shows strongly positive allochronical growth from 24 to 82 days EA (in S3–S8), a=2.92; between 87 days EA and birth, i.e., 280 days EA (in S9–S11), the positive trend decreases to a=1.45. The allochrony of the rostrum–eye distance (8), too, shows a decreasing trend: in 24–87 days EA (S3–S9),

Table 13. Selected dimensions of the head of *Stenella attenuata* (4, 5, 6, 8, 10, 11)

Stage	Range	Σ	*n*	x	SD	Sx
4. Head length						
3	2.5–3.6	12.7	4	3.2	0.4085	0.2042
4	3.8–5.5	36.8	8	4.6	0.6225	0.2200
5	6.2–9.5	131.4	17	7.7	0.8600	0.2086
6	8.2–14.2	162.2	14	11.6	1.9845	0.5303
7	18.9–24.5	217.8	10	21.8	2.0242	0.6401
8	24.4–33	243.7	9	27.1	2.9408	0.9802
9a	38–44	196	5	39.2	1.4805	1.4805
9b	45–47	186	4	46.5	0.4330	0.4330
9c	48–55	152	3	50.7	1.7848	1.7848
5. Rostrum–Nostrils						
3	0.3–0.4	1.5	4	0.4	0.0433	0.0216
4	0.4–1.4	5.4	8	0.7	0.3766	0.1332
5	1–2.7	33.2	17	1.9	0.6098	0.1479
6	3–6	61.7	14	4.4	0.9996	0.2671
7	8.3–12.2	102.9	10	10.3	1.1881	0.3757
8	12.2–16.2	123.7	9	13.7	1.2384	0.4128
9a	19–23	104	5	20.8	1.4697	0.6573
9b	26–28	107	4	26.8	0.8292	0.4146
9c	29–31	88	3	29.3	1.2472	0.7201
6. Rostrum–Ear						
3	0.6–0.8	3.0	4	0.7	0.1118	0.0559
4	1.0–2.0	11.9	8	1.5	0.4014	0.1419
5	2.4–4.2	55.8	17	3.3	0.5170	0.1254
6	3.7–8.3	81.4	14	5.8	1.5509	0.4145
7	11.7–16.5	146	10	14.6	1.6149	0.5107
8	17.3–24.2	168.5	9	18.7	2.1306	0.7102
9a	28–32	145	5	29	2.1902	0.9798
9b	33–37	140	4	35	1.4142	0.7071
9c	38–40	116	3	38.7	0.9428	0.5443
8. Rostrum–Eye						
3	0.7–1.0	3.5	4	0.9	0.1299	0.0649
4	0.9–1.7	11.8	8	1.5	0.2487	0.0879
5	2.2–3.2	45.3	17	2.7	0.3514	0.0852

Table 13. *continued*

Stage	Range	Σ	*n*	x	SD	Sx
S6	3.2–7.2	69.4	14	4.9	1.3113	0.3504
7	9.9–14	122.4	10	12.2	1.4305	0.4524
8	14.0–20	142.9	9	15.9	1.8594	0.6198
9a	23–26	119	5	23.8	1.1662	0.5215
9b	27–30	115	4	28.8	1.0897	0.5449
9c	31–34	97	3	32.3	1.2472	0.7201
10. Eye–Ear						
3	0.8–1.2	3.8	4	0.9	0.1915	0.0957
4	1.3–1.7	12.1	8	1.5	0.1642	0.0580
5	1.7–2.8	36.5	17	2.1	0.3125	0.0758
6	2.2–3.3	38.1	14	2.7	0.2914	0.0799
7	3.6–4.2	38.9	10	3.9	0.3143	0.0994
8	4.1–5.7	43.3	9	4.8	0.5395	0.1798
9a	6–7	33	5	6.6	0.5477	0.2449
9b	7–8	31	4	7.8	0.5000	0.2500
9c	8–10	26	3	8.7	1.1547	0.6667
11. Nostrils–Eye						
3	0.7–1.0	3.3	4	0.8	0.1258	0.0629
4	1.1–2.2	12.9	8	1.6	0.3720	0.1315
5	1.8–4	52.1	18	2.9	0.6690	0.1577
6	3.3–6.3	70.7	14	5.1	0.9882	0.2641
7	8.1–11	95.8	10	9.6	1.1726	0.3708
8	11.6–18	115.2	9	12.8	2.2294	0.7431
9a	16–19	88	5	17.5	1.1402	0.5099
9b	19–22	80	4	20	1.4142	0.7071
9c	22–22	66	3	22	0.0000	0.0000

x, mean; SD, standard deviation; Sx, standard error of the mean.

Fig. 20. Regression lines of allometric growth of head proportions in *S. attenuata*; logarithmic plotting. Ordinate: 6, rostrum–ear; 8, rostrum–eye; 10, eye–ear; abscissa: 2, total length (cm)

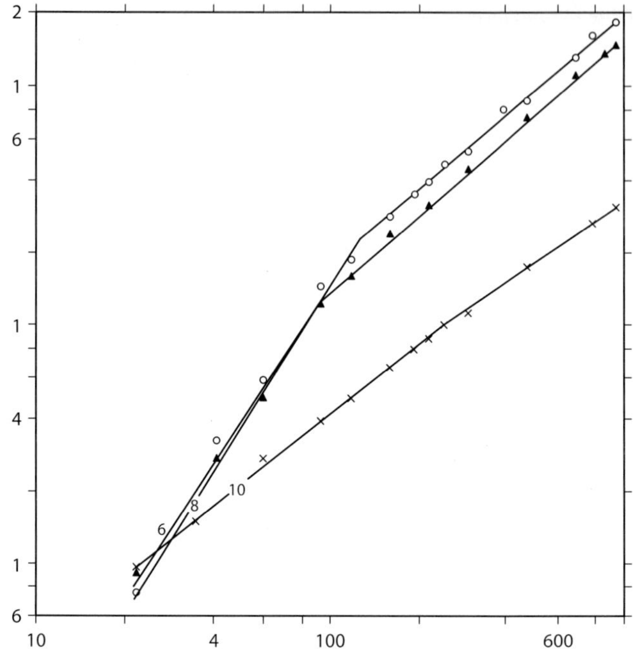

Fig. 21. Regression lines of allometric growth of head proportions in *S. attenuata*; logarithmic plotting. Ordinate: 5, rostrum–nostrils; 11, nostrils–eye; abscissa: 2, total length (cm)

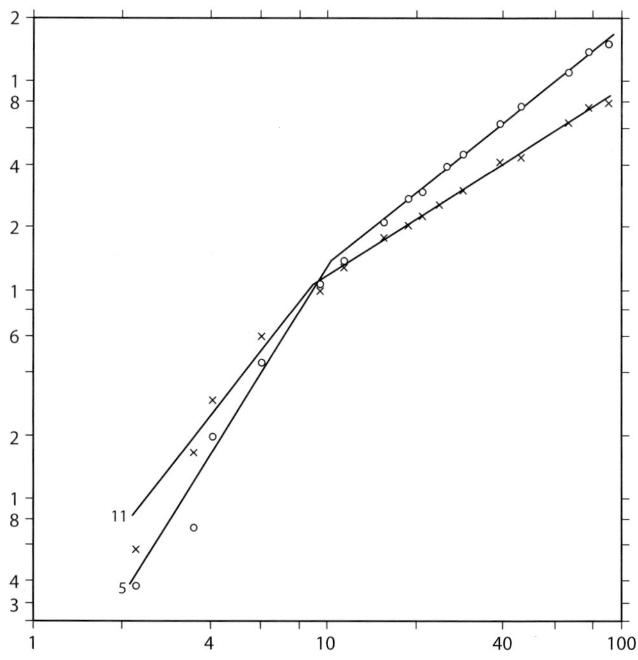

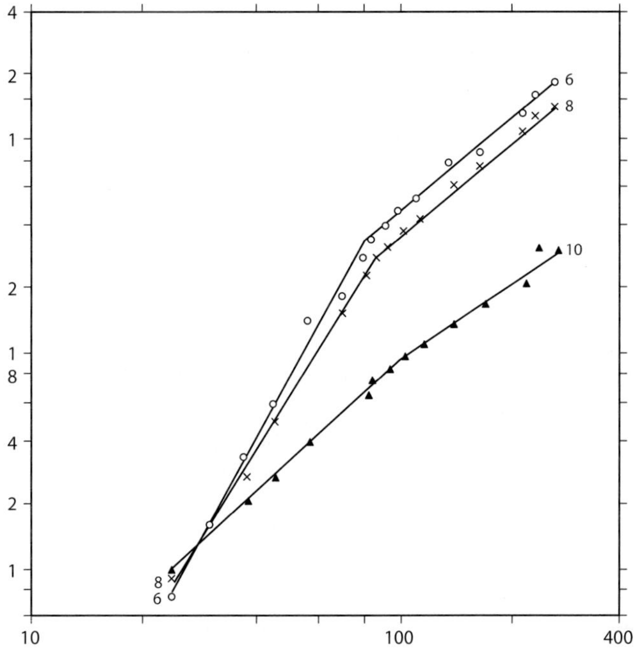

Fig. 22. Regression lines of allochronic growth of head proportions in *S. attenuata*; logarithmic plotting. Ordinate: 6, rostrum–ear; 8, rostrum–eye; 10, eye–ear; abscissa: time (days)

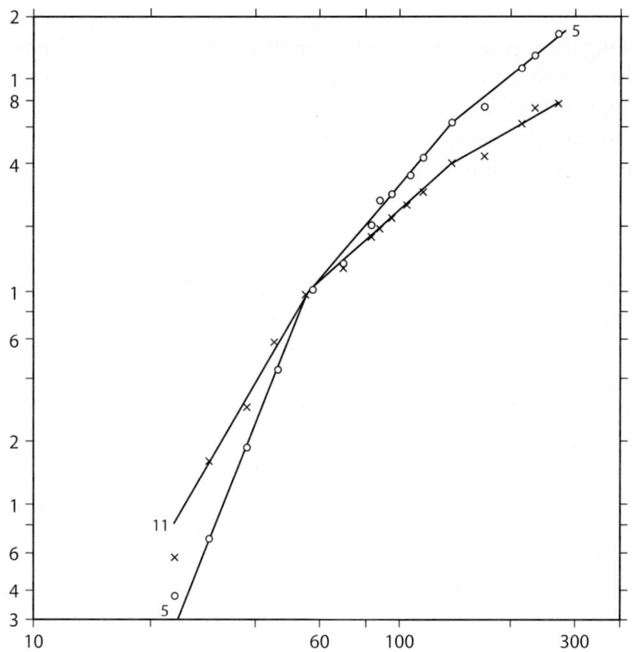

Fig. 23. Regression lines of allochronic growth of head proportions in *S. attenuata*; logarithmic plotting. Ordinate: 5, rostrum–nostrils; 11, nostrils–eye; abscissa: time (days)

a=2.70; in 95–280 days EA (S9–S11), this value decreases to a=1.43. However, the growth of the eye–ear distance (10) in time is positively allochronical: in 24–100 days EA (S3–S12), a=1.54; in 105–270 days EA (S10–S11), a=1.17 (Fig. 22). The allochronical growth of the rostrum–nostrils dimension (5) in 24–57 days EA (S3–S7), is strongly positive, a=4.2; from 57 to 140 days EA (S7–S10), a=2.0; from 140 to 280 days EA (from S10 to birth), it drops to a=1.44. The relation of the nostrils–eye dimension (11) shows a decreasing tendency from positively allochronical growth to isochronical: in 24–57 days EA (S3–S7), a=2.9; in 57–140 days EA (S7–S10), a=1.58; in 140–280 days EA (S10 until birth), a=1.0 (Fig. 23).

5.4 Development of the Thoracic Limb (Flipper)

At stage S3, the thoracic limb bud shows the usual shape as in other mammals. In S4 a mesenchymal condensation is visible in the longitudinal axis of the bud and a sinus marginalis forms in the distal margin of the enlarged handplate. The whole bud is 1.8 mm long and 1.3 mm wide at base. In S5 the chondral anlagen of the humerus, radius, and ulna are observed and characteristic non-segmented mesenchymal rays of metacarpal bones and digital phalanges are observed in the handplate. The indentations have not yet appeared between them. The handplate is symmetrical along the third ray. The bud is 2.5 mm long and 2 mm wide. At the beginning of stage 6, a fundamental change takes place in that the handplate changes into a flipper. In the prospective carpal joint the autopodium is abducted in the ulnar direction, the axis of the handplate is shifted from the third to the second ray which increases in length. The elongation is accompanied by multiplication of digital phalanges in the second and third digit. The anlagen of the carpal and metacarpal bones and digital phalanges chondrify. Older embryos in this stage show the onset of ossification of the scapula and clavicle (Sedmera et al. 1997b). The mean limb length at this stage is 4.1 mm and its mean width is 2.3 mm.

In the subsequent stages the flipper grows and its inner structure develops. In S9, ossification of humerus, radius, and ulna takes place. In the fetus Sa 75 (S9) six phalanges were demonstrated in the second and third digit, four phalanges in the fourth, and two phalanges in the first and fifth digits. In the largest fetus Sa 90, the length of the flipper is 102 mm and the width 52 mm.

Schildger (1986) studied the relations between the length and width of the flipper (Table 6, dimensions 13 and 14; Table 14). The growth of the length of the flipper (13) in relation to TL is in S3–S9 positively allometrical, a=1.25; between S9 and S11, a=1.04. The growth of the width of the flipper (14) is slightly positively allometrical; in S3–S9, a=1.18; in S10–S11, a=1.06 (Fig. 24). In relation to time, the growth of flipper length is positively allochronical: in 24–95 days EA (S3–S9), a=2.12; between 95–280 days EA (S9 to birth), a=1.54. The relative growth of the flipper width is positively allochronical: in 24–95 days EA (S3–S9), a=1.85; in 95–280 days EA (S9–birth), a=1.48 (Fig. 25).

Table 14. Selected dimensions of limbs of *Stenella attenuata* (13, 14, 19, 20)

Stage	Range	Σ	*n*	x	SD	Sx
13. Length of flipper						
3	1.1–1.7	7.0	5	1.4	0.2280	0.1019
4	1.4–2.0	12.5	8	1.6	0.1798	0.0635
5	1.4–3.0	43.6	18	2.4	0.4210	0.0992
6	3.2–5.2	55.6	14	4	0.6441	0.1721
7	6.3–9.3	78.1	10	7.8	0.9027	0.2855
8	9.3–11.7	87.9	9	9.8	0.9297	0.3099
9a	14–15	71	5	14.2	0.4000	0.1789
9b	17–19	70	4	17.5	1.1180	0.5590
9c	17–21	58	3	19.3	1.6997	0.9813
14. Width of flipper						
3	0.6–1.0	4.5	5	0.9	0.1549	0.0693
4	1.1–1.5	10.2	8	1.3	0.1561	0.0552
5	1.3–2.3	34.5	18	1.9	0.2794	0.0658
6	1.8–2.9	32.7	14	2.3	0.2990	0.0799
7	3.3–5.0	40.9	10	4.1	0.5049	0.1597
8	4.6–6.6	46.2	9	5.1	0.6074	0.2024
9a	7–9	36	5	7.2	0.4000	0.1789
9b	9–9	36	4	9	0.0000	0.0000
9c	10–11	32	3	10.7	0.4714	0.2722
19. Length of pelvic limb bud						
3	0.5–0.7	2.9	5	0.6	0.0748	0.0334
4	0.5–0.8	4.3	7	0.6	0.0989	0.0374
5	0.1–0.5	5	18	0.28	0.1133	0.0267
6	0.2–0.3	1.9	9	0.21	0.0314	0.0104
7						
8						
9a						
9b						
9c						
20. Width of pelvic limb bud						
3	0.4–0.4	0.4	1	0.4	0	0
4	0.4–0.8	2.4	4	0.6	0.1581	0.0790

Table 14. *continued*

Stage	Range	Σ	*n*	x	SD	Sx
6	0.2–0.3	1.8	8	0.22	0.0433	0.0153
7						
8						
9a						
9b						
9c						

x, mean; SD, standard deviation; Sx, standard error of the mean.

Fig. 24. Regression lines of allometric growth of flipper length and width in *S. attenuata*; logarithmic plotting. Ordinate: 13, length; 14, width; abscissa: 2, total length (cm)

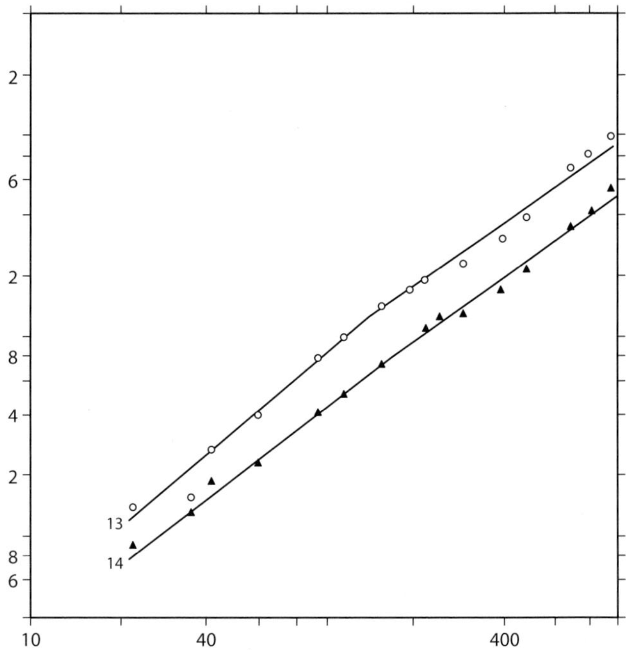

Fig. 25. Regression lines of allochronic growth of flipper length and width in *S. attenuata*; logarithmic plotting. Ordinate: 13, length; 14, width; abscissa: time (days)

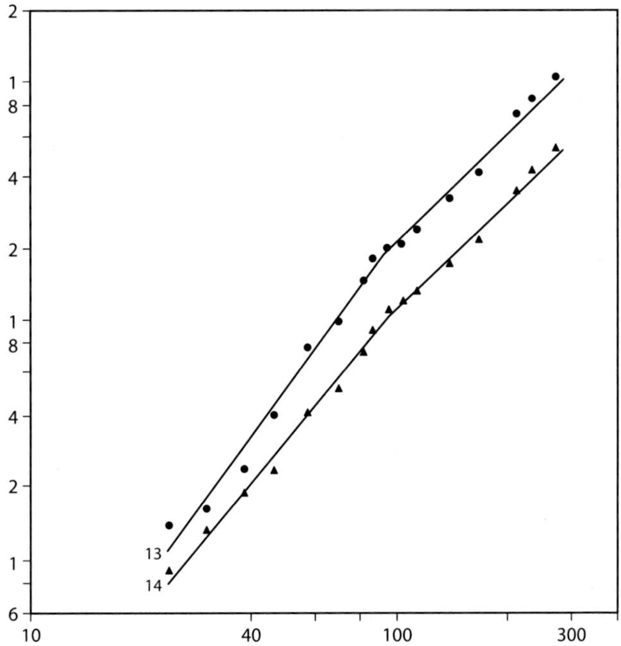

5.5 Development of Pelvic Limb Buds

Pelvic limb bud anlagen occur in all specimens examined at S3. They are approximately half the size of the thoracic limb buds and are filled with dense non-differentiated mesenchyme. They retain the same size even at S4 when suggestions of sinus marginalis and ingrowth of the spinal nerve into the mesenchyme are observed. In S5, the length of the buds drops to half, the buds become flattened and their distal ends are pointed. In cross-section, one can distinguish anlagen of the pelvic cartilage, head of femur, and blastema of prospective pelvic muscles. Under larger magnification, numerous apoptotic cells and macrophages can be observed (Sedmera et al. 1997a). The diminishing process continues even at stage 6 where, as stated by Schildger (1986), pointed tubercles 0.2 mm high could be observed in nine out of 14 embryos examined (Sa 24, Sa 32, Sa 35, Sa 36, Sa 37, Sa 38, Sa 39, Sa 40), and then only in embryo Sa 45 .Microscopically, it appears as a skin fold with a thicker epidermis and a more abundant vascular plexus. From embryo Sa 46 on, there are no traces of the pelvic limb buds and any remains of limb structures whatsoever have disappeared as well (Table 6, dimensions 19 and 20; Table 14).

5.6 Development of the Fluke

The first anlage of a fluke appears in three out of 18 embryos (Tables 6, 15) examined at S5 (Sa 20, Sa 27, Sa 30), forming elongated folds like isosceles triangles on either side of the tip of the tail bud. The base of the triangle (i.e., the length of fluke anlage) is 3 mm, and the height of the triangle (i.e., width of fluke anlage) amounts to 1.3 mm. In S6, fluke anlagen are found in nine out of 14 embryos (Sa 35, Sa 37, Sa 40–Sa 46), the triangular projections being 5 mm long and 2 mm wide (see Fig. 26).

Rudimentary flukes regularly occur upward of embryo Sa 40 (TL 64 mm), in which pelvic limb buds have regularly occurred for the last time. The fluke, still rhomboid in dorsal view, continues to grow in the course of S7 and S8, at which stage its caudal edge forms an arch with caudal convexity. At S6 and S7, the fluke is longer than wide; the two dimensions become equal at S8, and after the end of S8, its width is greater than its length. At S9 its lateral lobes increase in size (from Sa 68 on) and the caudal edge of the fluke changes from convex to convex–concave, with a median incision, attaining almost the definite shape in Sa 87 (S11), at which specimen the tips of the lateral lobes become pointed (Fig. 26).

Schildger (1986) studied the relative growth of the width of the fluke (16) in relation to TL. In S5–S12, a=2.0; in S12–S11, a=1.17. The growth of the length of the fluke (15) in S5–S12 is nearly isometrical, a=1.1; in S12–S11 positively allometrical, a=1.6 (Fig. 27). Štěrba et al. (1994) studied the growth of the fluke in relation to time. The width grows in 38–95 days EA (S5–S9) strongly positively allochronically, a=3.40; in 95–280 days EA (S9–birth), the growth slows to a=1.46. Also, the length of the fluke grows positively allochronically; from 38 to 95 days EA (S5–S9), a=2.0; in 95–280 days EA (S10–birth), a=1.4 (Fig. 28).

Table 15. Selected dimensions of appendages of *Stenella attenuata* (15, 16, 17, 18)

Stage	Range	Σ	*n*	x	SD	Sx
15. Length of fluke						
3						
4						
5	3.0–3.4	9.6	3	3.2	0.1633	0.0943
6	3.0–5.5	43.5	9	4.8	0.7165	0.2388
7	5.9–9.0	73.9	10	7.4	0.8859	0.2801
8	8.2–11.1	85.1	9	9.5	0.8098	0.2699
9a	13–16	70	5	14	1.0954	0.4899
9b	15–21	72	4	18	2.2361	1.1180
9c	19–20	59	3	19.7	0.4714	0.2722

Table 15. *continued*

Stage	Range	Σ	*n*	x	SD	Sx
16. Width of fluke						
3						
4						
5	1.2–1.4	3.9	3	1.3	0.0816	0.0471
6	1.6–2.3	17.7	9	2.	0.2403	0.0801
7	2.4–6.5	43	10	4.3	1.1099	0.3510
8	4.5–13.7	73	9	8.1	2.6946	0.8982
9a	17–23	97	5	19.4	2.0591	0.9209
9b	20–30	103	4	25.8	4.3803	2.1901
9c	31–37	102	3	34	2.4495	1.4142
17. Length of dorsal fin						
3						
4						
5						
6	6–6	12	2	6	0.0000	0.0000
7	9–12	102	10	10.2	0.0770	0.3406
8	10–16	113	9	12.6	1.9499	0.6499
9a	16–17	81	5	16.2	0.4000	0.1789
9b	20–23	83	4	20.8	1.2990	0.6495
9c	23–24	70	3	23.3	0.4714	0.2722
18. Height of dorsal fin						
3						
5	0.3	0.3	1	0.3	0.0000	0.0000
6	0.2-0.2	0.3	3	0.2	0.0000	0.0000
7	0.3–0.7	5	10	0.5	0.0894	0.0282
8	0.5–2.5	11.3	9	1.3	0.7588	0.2529
9a	3.0–5.0	20	5	4	0.8944	0.3999
9b	4–7	24	4	6	1.2247	0.6124
9c	8–7	25	3	8.3	0.4714	0.2722

x, mean; SD, standard deviation; Sx, standard error of the mean.

Fig. 26. Development of the fluke shape (right half) in *S. attenuata* in developmental stages S6–S11

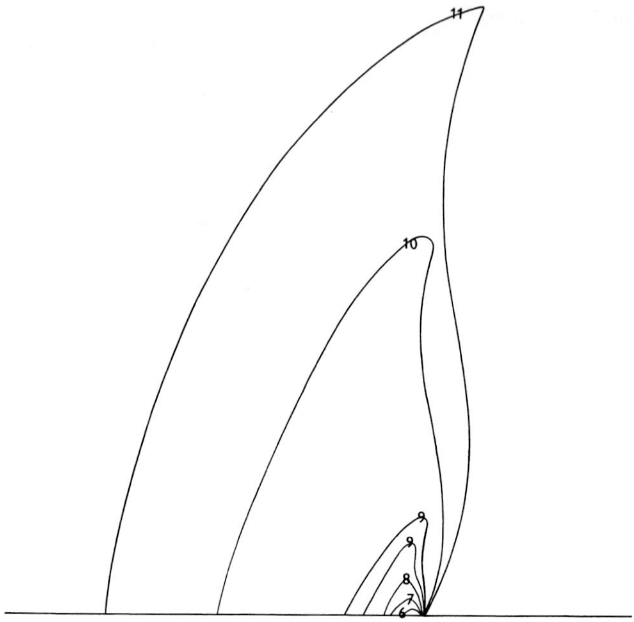

Fig. 27. Regression lines of allometric growth of the fluke in *S. attenuata*; logarithmic plotting. Ordinate: 15, length; 16,width of fluke; abscissa: 2, total length (cm)

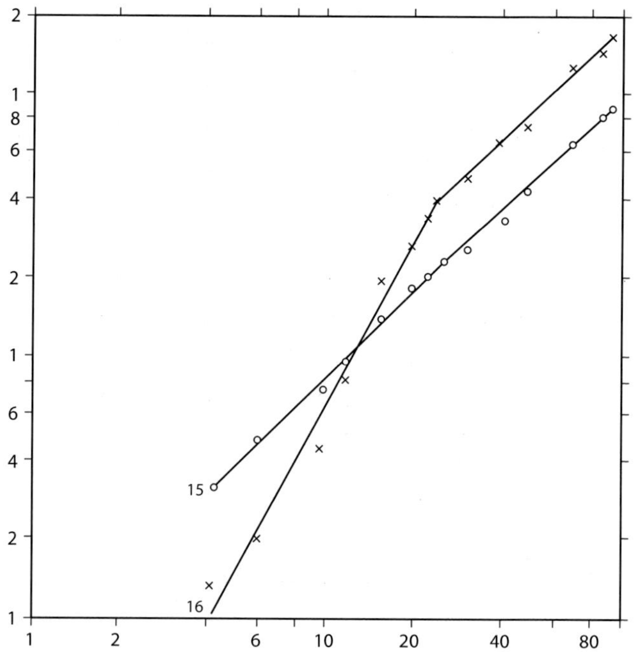

Fig. 28. Regression lines of allochronic growth of the fluke of *S. attenuata*; logarithmic plotting. Ordinate: 15, length; 16, width of fluke; abscissa: time (days)

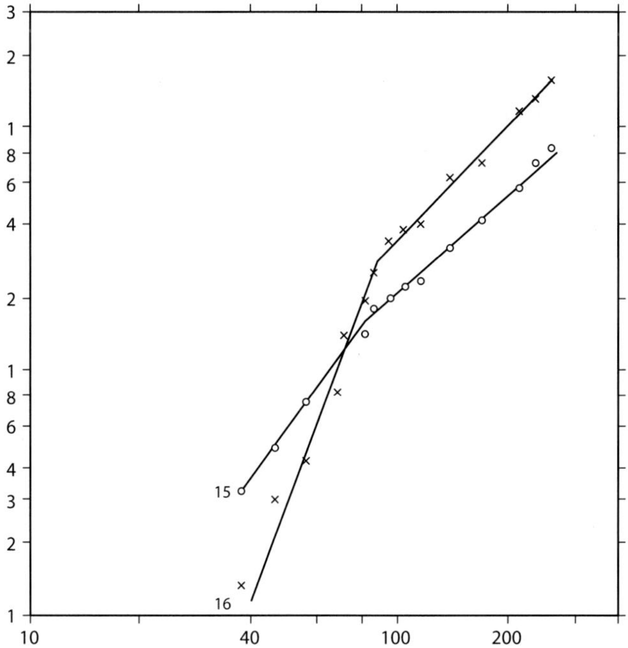

5.7 Development of the Dorsal Fin

For the first time, a dorsal fin appears in an older embryo at S5 (Sa 33, TL=44 mm) in the form of an elongate tubercle in the caudal part of the dorsum, 0.3 mm in height. At S6, we have found similar rudimentary dorsal fins in three out of 15 embryos (Sa 44–Sa 46); they attained the height of 0.2 mm and the length of 6 mm. From S7 on, a dorsal fin occurred in all fetuses, gradually increasing in size. Its growth accelerates at S9 when it attains from 17 to 24 mm in length and from 3 to 9 mm in height. In fetuses from Sa 77 to Sa 82 (S9, S12, S10), the dorsal fin is already pointed and is 23 mm

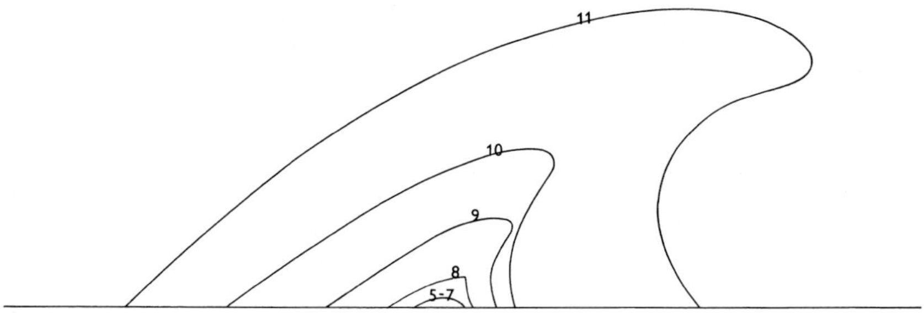

Fig. 29. The development of the shape of dorsal fin in *S. attenuata* in developmental stages S5–S11

Fig. 30. Regression lines of allometric growth of the dorsal fin in *S. attenuata*; logarithmic plotting. Ordinate: 17, length; 18, height of dorsal fin; abscissa: 2, total length (cm)

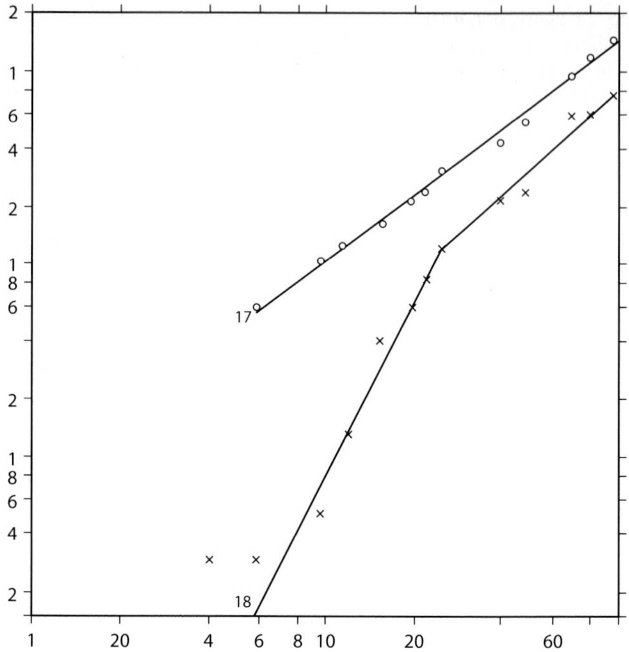

Fig. 31. Regression lines of allochronic growth of the dorsal fin in *S. attenuata*; logarithmic plotting. Ordinate: 17, length; 18, height of dorsal fin; abscissa: time (days)

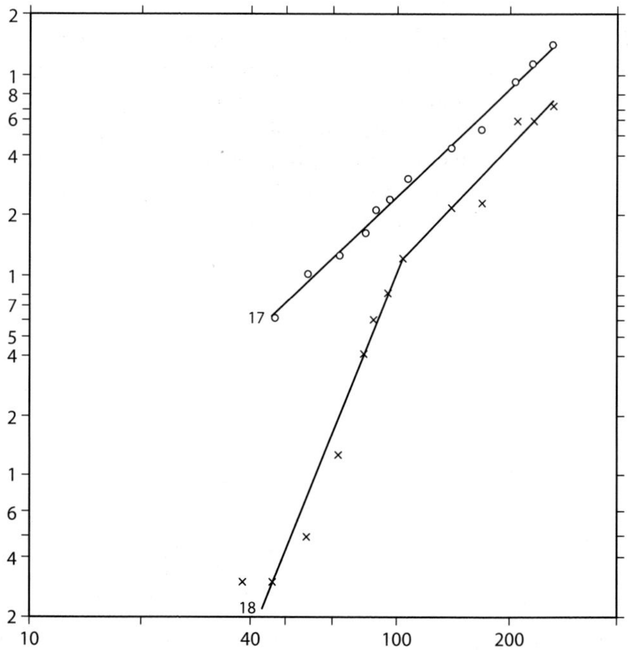

long and 8 mm high on average (Fig. 29). From S10 on (Sa 83, TL about 400 mm), the dorsal fin shows the species-specific shape and is 42 mm long and 21 mm high. The dimensions of the dorsal fin are given in Table 6 [fin height (17) and fin length (18)] and in Table 15.

Schildger (1986) studied the growth in height (18)and length (17) of the dorsal fin in relation to TL. Height: in S6–S12 the slope of the regression line a=3.00; in S12–S11, a=1.33. Length: in S6–S11 the slope of the regression line a=1.13 (Fig. 30). Štěrba et al. (1994) studied the growth of the dorsal fin height (18) and length (17) in relation to time. Height: in 46-104 days EA (S6–S12), a=4.59; in 105–280 days (S12–birth), a=1.93. Length also grows positively allochronically: in 46–280 days EA (S6–S11, birth), the slope of regression line, a=1.72 (Fig. 31).

5.8 Duration of Intrauterine Development

The duration of gestation and thus also intrauterine development of the four species under study cannot be different, as the adult females do not differ much in size. In *Stenella attenuata, S. longirostris,* and *Delphinus delphis* the body length varies between 1700 and 2000 mm and the weight between 80 and 100 kg; in *Phocoena phocoena,* between 1200 and 1800 mm and 50 and 75 kg, respectively. The data available on the lengths and weights of their newborns are scanty and it appears that they are not always original. The lengths of the newborns of the first three species vary between 750 and 1060 mm, their weights between 7 and 10 kg. The lengths of newborn *Phocoena phocoena* vary between 650 and 900 mm, their weights between 5 and 7.5 kg. Thus we are justified to assume that the intrauterine development of *Stenella attenuata, S. longirostris,* and *Delphinus delphis* will last equally long. Based on our calculations of the regression curves of proportional growth of various body parts in time, we have concluded that the intrauterine development of the above species will not last longer than 10 months, most probably 280±10 days, i.e., 270–290 days (see Tables 16, 17 for *Stenella attenuata;* Tables 18, 19 for *Stenella longirostris;* and Tables 20, 21 for *Delphinus delphis*). In the case of *Phocoena phocoena,* which is a little smaller, we assume roughly the same duration of intrauterine development. At any rate, it cannot last longer, since smaller species cannot show longer intrauterine development unless it is prolonged by delayed ovulation, delayed fertilization, or delayed implantation mechanisms.

Table 16. Estimated age of staged embryos and fetuses of *Stenella attenuata* (stages S1–S9). Presumed mean duration of intrauterine development 280±10 days. Ageing of S9 fetuses from total length and weight

Stage	Estimated age (days)		Total length (mm)		Weight (g)	
	270–290	280	Range	Mean	Range	Mean
S1	–	13	–	–	–	–
S2	–	17	–	–	–	–
S3	22–28	24	21–25	22	0.035–0.09	0.065
S4	27–36	30	28–39	35	0.079–0.22	0.14
S5	32–42	38	33–45	41	0.19–0.62	0.45
S6	41–52	46	42–74	60	0.61–2.2	1.29
S7	51–66	57	76–110	96	3.7–9.2	6.73
S8	62–78	70	112–134	119	11.5–21.4	15.1
S9	78–85	82	151–167	159	26.1–44.5	37.5
	85–90	87	187–205	194	52.2–64	56.4
	90–100	95	213–225	218	81–93	85

Table 17. Estimated age of fetuses of *S. attenuata* in stages S12, S10, and S11, from total length and weight

Stage	Range of total length	Fetus number	Total length (mm)	Weight (g)	Estimated age (days)	
					270–290	280
S12	231–250	Sa 80	245	133	102–108	104
S10	251–300	Sa 81	260	138	105–110	105
		Sa 82	295	178	110–125	115
	301–400	Sa 83	394	397	135–150	140
	401–500	Sa 84	468	693	160–175	170
		Sa 85	477	719	160–175	170
S11	501–600	–	–	–	–	–
	601–700	Sa 86	615	1525	190–210	195
		Sa 87	660	1945	200–220	205
		Sa 88	690	2105	210–230	215
	701–800	Sa 89	795	3225	220–240	235
	801–900	–	–	–	230–270	–
	901–1000	Sa 90	945	6950	260–290	270
S13 (N)		950		270–290	280	

Table 18. Estimated age of staged embryos and fetuses of *Stenella longirostris* (stages S1–S9). Presumed duration of intrauterine development 280±10 days. Ageing of S9 fetuses from total length and weight

Stage	Estimated age (days)		Total length (mm)		Weight (g)	
	270–290	280	Range	Mean	Range	Mean
S1	–	13	–	–	–	–
S2	–	17	–	–	–	–
S3	22–28	24	22–26	24	0.027–0.033	0.03
S4	27–36	30	31–38	34	0.09–0.18	0.13
S5	32–42	38	40–58	50	0.29–1.1	0.74
S6	41–52	46	66–79	75	1.65–3.5	2.6
S7	51–66	57	88–97	93	4.9–6.7	5.7
S8	62–75	70	114–137	130	10.7–23.9	17.6
S91	85		177		50	
	90		198		90	

Table 19. Estimated age of fetuses of *S. longirostris* in stages S12, S10, and S11 from total length and weight

Stage	Range of total length (mm)	Fetus number	Total length (mm)	Weight (g)	EA (days)	
					270–290	280
S12	201–230	Sl 27	215	89	95–105	95
S102	231–300	Sl 28	244	113	100–130	100
	301–400	Sl 29	320	–	130–160	135
	401–500	–	–	–	–	–
S113	501–600	Sl 30	590	–	190–220	210
	601–700	Sl 31	700	–	220–245	245
	701–800	Sl 32	720	–	245–270	250
	801–900	Sl 33	805	–	270–290	280
S13 (N)		800		280		

Table 20. Estimated age of staged embryos and fetuses of *Delphinus delphis* (stages S1–S9). Presumed duration of intrauterine development 280±10 days. Ageing of S9 fetuses from total length and weight

Stage	EA (days)		Total length (mm)		Weight (g)	
	270–290	280	Range	Mean	Range	Mean
S1	–	13	–	–		
S2	–	17	–	–		
S3	22–28	24	24–31	28	0.029–0.041	0.035
S4	27–36	30	32–39	35	0.12–0.16	0.14
S5	32–42	38	42–46	44	0.52–0.86	0.69
S6	41–52	46	52–69	61	0.82–1.7	1.31
S7	51–66	57	82–101	89	3.3–5.5	4.03
S8	62–72	66	114–144	129	10.6–22.3	15.4
S91	78–85	82	163–164	163	32.8–36.8	35
	85	194	–	55	–	

Table 21. Estimated age of fetuses of *D. delphis* in stages S12, S10, and S11, from total length and weight

Stage	Range of total length (mm)	Fetus number	Total length (mm)	Weight) (g)	Estimated age (days)	
					270–290	280
S12	200–250	Dd 23	237	91	85–100	90
S10	251–300	Dd 24	298	231	110–120	115
		Dd 25	300	456	135–145	140
	301–400	Dd 26	320	394	130–140	135
	401–500	Dd 27	482	854	160–175	165
S11	501–600	Dd 28	577	1250	175–195	185
	601–700	–	–	–	–	–
	701–800	–	–	–	–	–
	801–900	Dd 29	830	4010	250–270	255
S13 (N)			850	280		

5.9 Mating and Calving Seasons

Using the staging and ageing method, we determined the ontogenetic age of the embryos and fetuses and subtracted it from the date at which the mother had been captured. In this way we estimated the probable day of fertilization and, also, we were able to calculate the expected month of birth. We have found that populations of *Stenella attenuata, S. longirostris,* and *Delphinus delphis*, inhabiting the Tropics and Subtropics of the Pacific Ocean, will breed throughout the year. Their mating is more frequent from September to March, culminating from November to January. The young are born throughout the year, more frequently from June to December, most of births falling on August to October (see Fig. 32; Tables 22–24).

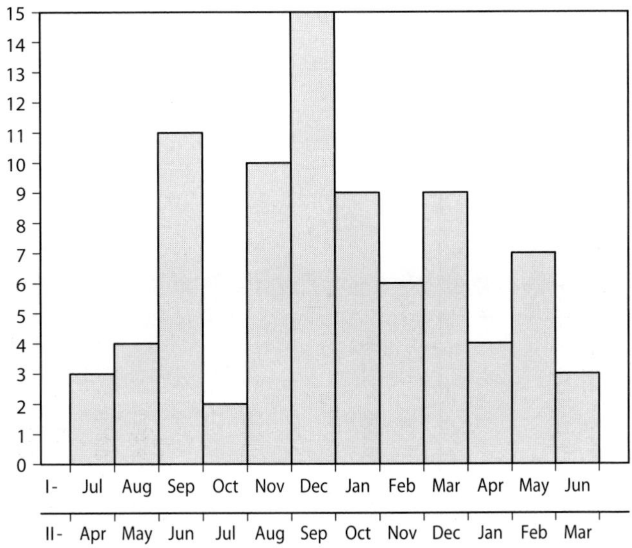

Fig. 32. Estimated numbers of fertilizations and of parturitions in *S. attenuata* in months of the year. Ordinate: numbers of fertilizations; abscissa I: months of presumed fertilizations; abscissa II: months of presumed parturitions

Table 22. Presumed months of fertilizations of females from which the embryos and fetuses under study were taken. *Stenella attenuata* (Sa)

Stage	Embryo number	TL (mm)	CRL (mm)	Weight (g)	EA (days) of embryos at presumed pregnancy		Date of catching of mother	Estimated month of conception at presumed duration of pregnancy		
					Range 270–290	Mean 280		270 days	290 days	280 days
S3	Sa 2	–	–	0.085	22–28	24	29.04.73	April	April	April
	Sa 3	21	8	0.035			22.10.72	Oct.	Sept.	Sept.
	Sa 4	21	10	0.090			25.02.74	Feb.	Jan.	Feb.
	Sa 5	22	9	0.051			06.04.75	March	March	March
	Sa 6	25	9	0.062			30.04.73	April	April	April
S4	Sa 7	28	11	0.081	27–36	30	28.02.74	Feb	Jan.	Jan.
	Sa 8	32	11	0.089			29.04.73	April	March	April
	Sa 9	33	11	0.093			07.02.73	Jan.	Jan.	Jan.
	Sa 10	–	11	0.079			24.04.73	March	March	March
	Sa 12	35	12	0.16			29.04.73	April	March	March
	Sa 14	38	13	0.20			12.06.71	May	May	May
	Sa 15	38	13	0.22			03.02.73	Jan.	Dec.	Jan.
	Sa 16	39	12	0.19			07.02.73	Jan.	Jan.	Jan.
S5	Sa 11	33	14	0.19	32–42	38	25.01.	Dec.	Dec.	Dec.
	Sa 13	35	15	0.28			14.06.79	May	May	May
	Sa 17	39	15	0.30			04.02.74	Jan.	Dec.	Dec.
	Sa 18	39	15	0.30			10.01.75	Dec.	Nov.	Dec.

Table 22. *continued*

Stage	Embryo number	TL (mm)	CRL (mm)	Weight (g)	EA (days) of embryos at presumed pregnancy		Date of catching of mother	Estimated month of conception at presumed duration of pregnancy		
					Range 270–290	Mean 280		270 days	290 days	280 days
	Sa 19	–	15	0.28			24.09.78	Aug.	Aug.	Aug.
	Sa 20	41	17	0.50			15.02.76	Jan.	Jan.	Jan.
	Sa 21	41	18	0.49			10.10.79	Sept.	Aug.	Sept.
	Sa 22	41	21	0.49			24.06.75	May	May	May
	Sa 23	42	19	0.50			27.01.	Dec.	Dec.	Dec.
	Sa 25	43	17	0.47			15.10.79	Sept.	Sept.	Sept.
	Sa 26	43	17	0.44			21.10.79	Sept.	Sept.	Sept.
	Sa 27	43	17	0.54			30.10.79	Sept.	Sept.	Sept.
	Sa 28	43	18	0.45			23.01.73	Dec.	Dec.	Dec.
	Sa 30	43	19	0.50			06.11.63	Oct.	Sept.	Sept.
	Sa 31	44	19	0.57			08.02.74	Jan.	Dec.	Jan.
	Sa 33	44	19	0.56			19.01.74	Dec.	Dec.	Dec.
S6	Sa 24	42	22	0.71	41–52	46	21.04.75	March	March	March
	Sa 32	44	19	0.61			10.02.74	Dec.	Dec.	Dec.
	Sa 35	48	22	0.85			22.01.76	Dec.	Dec.	Dec.
	Sa 36	49	22	0.86			13.03.74	Feb.	Jan.	Jan.

Table 22. *continued*

Stage	Embryo number	TL (mm)	CRL (mm)	Weight (g)	EA (days) of embryos at presumed pregnancy		Date of catching of mother	Estimated month of conception at presumed duration of pregnancy		
					Range 270–290	Mean 280		270 days	290 days	280 days
	Sa 37	52	23	0.89			08.03.74	Jan.	Jan.	Jan.
	Sa 40	64	28	1.6			26.10.79	Sept.	Sept.	Sept.
	Sa 41	65	29	1.7			29.01.75	Dec.	Dec.	Dec.
	Sa 42	66	26	1.5			18.10.79	Sept.	Aug.	Sept.
	Sa 43	67	30	1.6			12.10.79	Sept.	Aug.	Aug.
	Sa 44	70	30	1.6			15.08.76	July	June	July
	Sa 45	71	31	1.7			22.01.75	Dec.	Nov.	Nov.
	Sa 47	74	30	–			13.02.73	Jan.	Dec.	Dec.
S7	Sa 48	76	35	–	51–66	57	02.02.76	Dec.	Nov.	Dec.
	Sa 49	85	37	3.7			08.11.74	Sept.	Sept.	Sept.
	Sa 50	89	43	4.7			08.01.72	Nov.	Nov.	Nov.
	Sa 51	91	44	5.0			13.11.79	Sept.	Sept.	Sept.
	Sa 52	92	47	4.1			25.04.72	March	Feb.	Feb.
	Sa 54	100	44	7.2			20.04.74	Feb.	Feb.	Feb.
	Sa 55	103	46	9.2			15.03.74	Jan.	Jan.	Jan.
	Sa 56	107	55	8.6			13.01.72	Nov.	Nov.	Nov.
	Sa 57	110	47	8.7			20.04.74	Feb.	Feb.	Feb.

Table 22. *continued*

Stage	Embryo number	TL (mm)	CRL (mm)	Weight (g)	EA (days) of embryos at presumed pregnancy		Date of catching of mother	Estimated month of conception at presumed duration of pregnancy		
					Range 270–290	Mean 280		270 days	290 days	280 days
	Sa 58	110	52	8.8			31.01.72	Dec.	Nov.	Dec.
S8	Sa 59	112	57	12.0	62–75	70	22.05.73	March	March	March
	Sa 60	112	63	13.3			06.02.75	Dec.	Nov.	Nov.
	Sa 61	114	58	12.5			31.01.73	Nov.	Nov.	Nov.
	Sa 62	114	61	12.5			13.02.74	Dec.	Nov.	Dec.
	Sa 63	115	59	11.5			26.04.79	Feb.	Feb.	Feb.
	Sa 64	116	46	17.7			13.02.74	Dec.	Nov.	Dec.
	Sa 65	120	62	18.6			20.04.74	Feb.	Feb.	Feb.
	Sa 66	133	61	16.8			17.05.74	March	March	March
	Sa 67	134	73	21.4			08.10.79	Aug.	July	July
S9	Sa 68	151	80	26.1	78–85*	82	14.06.79	March	March	March
	Sa 69	153	82	44.5			04.02.74	Nov.	Nov.	Nov.
	Sa 70	157	85	36.0			06.01.	Oct.	Oct.	Oct.
	Sa 71	165	86	36.5			25.02.74	Dec.	Nov.	Dec.
	Sa 72	167	93	44.5			30.01.75	Nov.	Oct.	Nov.
	Sa 73	187	101	64.0	85–90	87	27.06.75	April	March	April
	Sa 74	192	101	54.1			26.10.79	Aug.	July	Aug.

Table 22. *continued*

Stage	Embryo number	TL (mm)	CRL (mm)	Weight (g)	EA (days) of embryos at presumed pregnancy		Date of catching of mother	Estimated month of conception at presumed duration of pregnancy		
					Range 270–290	Mean 280		270 days	290 days	280 days
	Sa 75	193	101	52.2			27.05.79	March	Feb.	March
	Sa 76	205	108	55.3			10.10.78	July	June	June
	Sa 77	213	102	81	90–100	95	27.01.74	Nov.	Oct.	Nov.
	Sa 78	215	110	82			22.02.76	Dec.	Nov.	Nov.
	Sa 79	225	112	93			06.01.	Oct.	Oct.	Oct.
S12	Sa 80	245	132	133	102–108	104	08.03.78	Nov.	Nov.	Nov.
S10	Sa 81	261	143	138	105–110	105	26.06.75	March	March	March
	Sa 82	295	152	178	110–125	115	13.10.78	June	June	June
	Sa 83	394	200	397	135–150	140	11.10.72	May	May	May
	Sa 84	468	220	693	160–175	170	05.11.77	May	May	May
	Sa 85	477	230	719	160–175	170	05.11.77	May	May	May
S11	Sa 86	615	300	1525	190–210	195	30.03.75	Sept.	Sept.	Sept.
	Sa 87	660	340	1945	200–220	205	12.03.78	Aug.	Aug.	Aug.
	Sa 88	690	330	2105	210–230	215	28.02.80	July	July	July
	Sa 89	795	400	3225	220–240	235	11.01.79	May	May	May
	Sa 90	945	510	6950	260–290	270	16.03.79	June	May	June

* The age of fetuses in S9, S10, S11, and S12 was estimated from total length and weight.
TL, total length; CRL, crown–rump length; EA, estimated age.

Table 23. Presumed months of fertilizations of females from which the embryos and fetuses under study were taken. *Stenella longirostris* (Sl)

Stage	Embryo number	TL (mm)	CRL (mm)	Weight (g)	EA (days) of embryos at presumed of pregnancy		Date of catching of mothero	Estimated month of conception at presumed duration of pregnancy		
					Range 270–290	Mean 280		270 days	290 days	280 days
S 3	Sl 1	–	–	–	22–28	24	20.05.73	April	April	April
	Sl 2	22	8	0.033			01.07.79	June	June	June
	Sl 3	26	9	0.027			21.02.76	Jan.	Jan.	Jan.
S4	Sl 4	31	11	0.09	27–36	30	07.05.75	April	March	April
	Sl 5	33	12	0.13			20.01.74	Dec.	Dec.	Dec.
	Sl 6	34	12	0.12			27.02.76	Feb.	Jan.	Jan.
	Sl 7	38	13	0.18			03.02.73	Jan.	Dec.	Jan.
S5	Sl 8	40	15	0.29	32–42	38	30.04.72	March	March	March
	Sl 9	52	22	0.84			18.04.75	March	March	March
	Sl 10	58	23	1.1			06.01.74	Dec.	Nov.	Nov.
S6	Sl 11	66	27	1.65	41–52	46	26.10.79	Sept.	Sept.	Sept.
	Sl 12	72	28	2.1			18.06.73	May	April	May
	Sl 13	78	30	2.47			22.04.75	March	March	March
	Sl 14	78	35	3.5			20.02.76	Jan.	Dec.	Jan.
	Sl 15	79	33	3.15			18.06.73	May	April	May
S7	Sl 16	88	39	5.2	51–66	57	02.02.71	Dec.	Nov.	Dec.
	Sl 17	91	40	6.1			26.09.76	Aug.	July	Aug.
	Sl 18	96	45	6.7			18.04.75	Feb.	Feb.	Feb.

Table 23. *continued*

Stage	Embryo number	TL (mm)	CRL (mm)	Weight (g)	EA (days) of embryos at presumed of pregnancy		Date of catching of mother	Estimated month of conception at presumed duration of pregnancy		
					Range 270–290	Mean 280		270 days	290 days	280 days
	Sl 19	97	45	4.9			14.06.79	April	April	April
S8	Sl 20	114	53	10.7	62–75	70	22.04.75	Feb.	Feb.	Feb.
	Sl 21	134	65	23.9			06.12.71	Oct.	Sept.	Sept.
	Sl 22	136	66	19.6			18.04.75	Feb.	Feb.	Feb.
	Sl 23	137	63	16.8			20.04.75	Feb.	Feb.	Feb.
	Sl 24	–	67	17.1			02.02.71	Dec.	Nov.	Nov.
S9	Sl 25	177	93	50	68–78	75	21.04.75	Feb.	Feb.	Feb.
	Sl 26	198	103	90			19.01.75	Nov.	Nov.	Nov.
S12	Sl 27	215	107	89	95–105	95	23.01.74	Oct.	Oct.	Oct.
S10	Sl 28	244	108	113	100–130	100	15.03.74	Dec.	Nov.	Dec.
	Sl 29	320	137	–	130–160	135	03.02.79	Sept.	Aug.	Sept.

TL, total length; CRL, crown–rump length; EA, estimated age.

Table 24. Presumed months of fertilizations of females from which the embryos and fetuses under study were taken. *Delphinus delphis* (Dd)

Stage	Embryo number	TL (mm)	CRL) (mm)	Weight (g)	EA (days) of embryos at presumed pregnancy		Date of catching of mother	Estimated month of conception at presumed duration of pregnancy		
					Range 270–290	Mean 280		270 days	290 days	280 days
S4	Dd 3	32	13	0.12	27–36	30	27.01.74	Dec.	Dec.	Dec.
S6	Dd 8	52	22	0.82	41–52	43	26.03.79	Feb.	Feb.	Feb.
	Dd 9	54	24	1.01			30.04.74	March	March	March
	Dd 11	69	30	1.7			28.06.75	May	May	May
S7	Dd 13	84	35	3.3	51–65	52	24.05.75	April	March	April
S8	Dd 16	120	51	13.4	62–72	66	28.01.74	Nov.	Nov.	Nov.
	Dd 18	135	60	19.4			21.06.75	April	April	April
	Dd 19	144	56	22.3			03.04.79	Jan.	Jan.	Jan.
S9	Dd 22	194	77	55	83		08.02.74	Nov.	Nov.	Nov.
S10	Dd 24	298	128	231	110–120	115	02.04.79	Dec.	Dec.	Dec.
	Dd 25	300	165	456	135–145	140	23.02.75	Oct.	Oct.	Oct.
	Dd 26	320	172	394	130–140	135	30.06.76	Feb.	Feb.	Feb.

TL, total length; CRL, crown–rump length; EA, estimated age.

6 Discussion

In this part, we shall concentrate upon two major problem areas or objections that could be raised, viz., (1) the method used, particularly the part pertaining to the ageing, and (2) the duration of intrauterine development and estimation of the mating and calving seasons.

6.1 Comparison of the Staging and Ageing Method with Other Methods

The classification of embryos and fetuses in selected stages (staging) is based on the study of definite characteristics (external shape, discrete characters, size, weight). Embryos showing identical definite characteristics are classified in one and the same stage. Stages 1–6 comprise the embryonic development, stages 7–13 the fetal one. Thus the pertinence to a stage does not depend merely on size or on weight. During embryonic development, an expression of homeobox genes takes place, which are common to higher taxa, and thus the development of body shape is considerably uniform in all mammalian species, epigenetic mechanisms making themselves felt to a considerable degree in the small embryo.

Consequently, embryos belonging to the same stage are assumed to be of the same ontogenetic age. In embryonic stages 1–6, the ontogenetic age was read from regression curves (see Fig. 2, after Štěrba 1995). In our opinion, the embryonic period of the species under study lasts no longer than 45 days, and the TL of the embryos is at least 40 mm and their CRL 20 mm. Also, Yablokov et al. (1972) state that the embryonic period of cetaceans ends at 35–45 days, and that the embryo of *Delphinapterus leucas* at the age of 45 days will attain a length around 40 mm. Besides, the authors record several earlier attempts at determining the age of cetacean fetuses by the longitudinal growth of their body or by the increase in their weight (Yablokov et al. 1972). In his paper on fetal growth rates of whales, with special reference to the fin whale *Balaenoptera physalus* Linn., Laws (1959) based methodically on the paper by Huggett and Widdas (1951). These authors have demonstrated that the fetal length increases linearly with age, according to the formula: $L=a(t-t_0)$, where a is a constant denoted as the "specific fetal growth velocity", t is the duration of gestation period, and t_0 is the intercept where the linear part of the plot, if produced backwards, cuts the time axis; in other words, it expresses the duration of initial slow growth. Huggett and Widdas (1951)use arbitrary estimates of t_0: e.g., for gestation from 50 to 100 days, t_0=0.3x(gestation time); for gestations from 100 to 400 days, t_0=0.2x(gestation time). In our case,

0.2×280 days=56 days, which, in our opinion, is too many days for so-called slow growth.

Our earlier studies (Štěrba 1995) indicate that, contrary to what had been thought previously, the rapid growth of some dimensions starts much earlier, already in embryos in stages where their TL or CRL can be measured. The slow growth period, which in fact has been synonymous with the embryonic period, comprises only the periods of cleavage, gastrulation, and onset of notogenesis, that is, the beginning of the embryonic period.

The deficiency of those papers which ignore the degree of morphogenesis of the embryos is in that they merely study such quantitative indices as length or weight. Length or weight provides poor information on the actual degree of development, particularly as they may differ in the methods used to measure them, which in turn makes it difficult to compare data presented by different authors. We can see the advantage of the staging method in that the stages permit, particularly in the early period, to determine the degree of development, thus correcting the quantitative indices, and then to determine the age on that basis. In this way, one can obtain a more precise idea of the onset, end, and rates of different developmental events and of their mutual relation in time.

6.2 Duration of Intrauterine Development and the Mating and Calving Seasons

Data from the literature on duration of intrauterine development, summarized in the introduction of this paper, vary from 6 to 12 months, most frequently from 9 to 12 months. In our opinion, the data have been mostly obtained from observations at open sea, using observations of mating, findings of the first embryos and fetuses in captured females, and the occurrence of first newborns to conclude on the duration of intrauterine development. While we do not doubt on the significance of such observations for the knowledge of the biology of dolphins, it is our opinion that data on the duration of gestation, obtained in this way, cannot be but rough estimations. In our determinations of the gestation period and thus the intrauterine development, we have started from the data in the literature summed up in the introduction to this paper, stating that the adult females of the four species under study do not differ much in size. Also, the lengths of newborns of the first three species are very similar, varying between 750 and 1060 mm, their weights between 7 and 10 kg; the lengths of newborn *Phocoena phocoena* vary between 650 and 900 mm and their weights between 5 and 7.5 kg. Thus we are justified in concluding that the intrauterine development of *Stenella attenuata*, *S. longirostris*, and *Delphinus delphis* has the same duration. On the basis of calculations of the regression curves of the proportional growth of various body parts in time, we have concluded that the intrauterine development of these species will not last longer than 10 months, most probably 280±10 days, i.e., 270–290 days. For *Phocoena phocoena*, a somewhat smaller species, we assume roughly the same duration of intrauterine development, provided that it is not prolonged by delayed ovulation, delayed fertilization, or delayed implantation mechanisms. In most cases, the mating and calving seasons have been defined only vaguely but, nevertheless, they have been placed in certain seasons of the year (see data from the literature in the Introduction). Reproduction throughout the year has only been documented for *Delphinus delphis*. However, land mammals in tropical regions do not show defined mating seasons and they reproduce throughout the year. Applying the staging and ageing method to the embryos and fetuses of *Stenella attenuata, S. longirostris*, and *Delphinus delphis* in our material, coming from the tropical to subtropical zones of the Pacific ocean, we have determined their ontogenetic age, which we then subtracted from the date on which the mothers had been captured. In this way we have obtained the probable date of fertilization and we could calculate the month of expected birth. We have found that the above species will reproduce throughout year, yet we can nevertheless observe certain seasonality of these phenomena. Matings are more frequent from September to March, with a peak from November to January. The young are born throughout the year, too, with a corresponding higher frequency of births from June to December, most of them falling from August to October. The data in the literature pertaining to *Phocoena phocoena*, a species inhabiting the Temperate Zone of the Atlantic and the Baltic, are fairly uniform in this respect. This species will mate from spring until autumn, and births have also been observed roughly during this period, which is not at variance with the data on the gestation period lasting 9 months.

References

Asdell, S. A. (1946): Patterns of mammalian reproduction. (Comstock Publ.Co., New York)

Bryden, M. M. (1972): Growth and development of marine mammals. In: Harison, R. J. (ed.): Functional Anatomy of Marine Mammals, Vol. 1: 2–24, (Academic Press, London -New York).

Bryden, M. M., Harrison, R. J. (1986): Gonads and reproduction. In: Bryden, M. M. (ed.): Research on Dolphins, Chapter 9: 149–159, (Clarendon Press, Oxford).Buffrenil, V. De., Collet, A. (1983): Donnés méthodologiques sur l´emploi de la technique squelettochronologique chez le Dauphin commun (*Delphinus delphis* L.). Ann. Sci. Nat. Zool. Paris, 13: 269–285.

Buhl, E. H., Oelschläger, H. A. (1986): Ontogenetic development of the nervus terminalis in toothed whales. Evidence for its non -olfactory nature. Anat. Embryol., 173: 285–294.

Buhl, E. H., Oelschläger, H. A. (1988): Morphogenesis of the brain in the harbour porpoise. J. Comp. Neurol., 227: 190–225.

Collet, A. (1994): *Delphinus delphis*. In: Niethammer, J., Krapp, F. (eds.): Handbuch der Säugetiere Europas, 6. Meeressäuger: Robineau, D., Duguy, R., Klima, M. (eds.), (Aula - Verlag, Wiesbaden).

Evans, P. G. H. (1987): The natural history of whales and dolphins, (Facts on File Inc., New York - Oxford).

Guldberg, G., (1894): On the duration of gravidity in certain Odontoceti. Bergen Museum, 4: 61–65. Cited after Yablokov (1972).

Harrison, R. J., (1969): Reproduction and reproductive organs. In: Andersen, H. J. (ed.): The Biology of Marine Mammals. Chapter 8, 253–390, (Academic Press, New York - San Francisco - London).

Harrison, R. J., Brownell, R. L., Boice, R. C. (1972): Reproduction and gonadal appearances in some odontocetes. In: Harrison, R. J. (ed.): Functional Anatomy of Marine Mammals, Vol. 1, 397–419, (Academic Press, London - New York).

Huggett, A. St . G., Widdas, W. F. (1951): The relationship between mammalian foetal weight and conception age. J.Physiol., 114: 306–317.

Hui, C. A. (1979): Correlates of maturity in the common dolphin, *Delphinus delphis*. Fishery Bull., 77(1): 295–300.

Kasuya, T., Miyazaki, N., Dawbin, W. H. (1974): Growth and reproduction of *Stenella attenuata* in the Pacific coast of Japan. Sci. Rep. Whales Res. Inst., Tokyo, 26: 157 –226.

Kinze, C. C. (1994): *Phocoena phocoena*. In: Niethammer, J., Krapp, F.(eds.): Handbuch der Säugetiere Europas, 6. Meeressäuger: Robineau, D., Duguy, R., Klima, M. (eds.), (Aula-Verlag, Wiesbaden).

Kleinenberg, S. E., (1956): Mammals of the Black Sea and Azov Sea. (Mlekopitayuchtchye Tchernogo i Azovskogo morey), (Publ. House of Acad. of Sci. USSR, Moscow-Leningrad, in Russian).

Klima, M. (1978): Comparison of early development of sternum and clavicle in striped dolphin and in humpback whale. Sci. Rep. Whales Res. Inst., 30: 253–269.

Klima, M. (1990): Rudiments of the clavicle in the embryos of whales (Cetacea). Z. Säugetierkunde, 55: 202–212.

Klima,M.: (1999): Development of cetacean nasal skull. Advances in Anatomy, Embryology and Cell Biology, 149: 1 - 143. (Springer Verlag, Berlin-Heidelberg-New York).

Klima, M., van Bree, P. J. H. (1985): Überzählige Skelettelemente im Nasenschädel von *Phocoena phocoena* und die Entwicklung der Nasenregion bei den Zahnwalen. Morph. Jb., 131: 131–178.

Klima, M., van Bree, P. J. H. (1990): On the origin of the so-called Meckelian ossicles in the nasal skull of Odontocetes. Morph. Jb., 136: 431–434.

Laws, R. M. (1959): The fetal growth rates of whales with special reference to the fin whale, *Balaenoptera physalus* Linn., Discovery Reports, 29: 281–308.
Leatherwood, S., Reeves, R. R. (1983): Whales and dolphins. (Sierra Club Books, San Francisco).
Lick, R. R. (1987): Untersuchungen zur Embryonalentwicklung des Integuments bei Cetaceen (*Physeter macrocephalus, Balaenoptera acutorostra, Phocoena phocoena).* Diplomarbeit, Zentrum f. Morphologie, J.- W. – Goethe – Universität, Frankfurt / M.
Martin, A. R. (1992): Das große Buch der Wale und Delphine. (Kremayr/Scherian und Gutersloh, Bertelsmann, Wien).
Mc Bride, A. F., Kritzler, H. (1951): Observations on pregnancy, parturition and postnatal behaviour in the bottlenose dolphin, J. Mammal., 32: 251–266.
Mead, J. G., Odell, D. K., Wells, R. S., Scott, M. D. (1980): Observations of a mass stranding of spinner dolphin, *Stenella longirostris,* from the West coast of Florida. Fish Bull. U.S. 78: 353–360.
Meyer, W., Neurand, K., Klima, M. (1995): Prenatal development of the integument in Delphinidae (Cetacea: Odontoceti). J. Morphol., 223: 269–287.
Míšek, I., Witter, K., Peterka, M., Lesot, H., Štěrba, O., Klima, M., Tichý, F., Peterková, R. (1996): Initial period of tooth development in dolphins *(Stenella attenuata,* Cetacea) – A pilot study. Acta vet. Brno, 65: 277–284.
Moehl-Hansen, U. (1954): Investigation on reproduction and growth of the porpoise (*Ph. phocoena* L.) of the Baltic. Vidensk. Medd. Dansk. Natur. Foren. Kobenhavn, 116: 369–396.
Nishiwaki, M. (1972): General biology. In: Ridgway, S. H. (ed): Mammals of the sea. Biology and medicine. (C.C. Thomas Publ., Springfield, Illinois).
Nishiwaki, M., Kasuya, T., Tobayana, T., Kamiya, T., Nakajima, M. (1965): *Feresa attenuata* were caught and kept at Ito, Japan. Scient. Rep. Whales Res. Inst. Tokyo, 19: 65–90.
Norris, K. S. (ed.) (1966): Whales, Dolphins and Porpoises. (Univ. Calif. Press, Berkeley – Los Angeles).
Nowak, R. M., Paradiso, J. L. (1983): Walker's Mammals of the World, 4th ed., Vol. 2, (Johns Hopkins University Press, Baltimore).
Oelschläger, H. A., Buhl, E. H. (1985): Development and rudimentation of the peripheral olfactory system in the harbor porpoise *Phocoena phocoena* (Mammalia:Cetacea). J. Morph., 184: 351–360.
Oelschläger, H. A., Buhl, E. H. (1987): Development of the Nervus terminalis in mammals including toothed whales and humans. In: Demski, L. S., Schwanzel-Fukunda, M., (eds.): The Terminal Nerve (Nervus terminalis). Structure, Function and Evolution. Ann. NY Acad. Sci., Vol. 519.
Ohsumi, S. (1966): Allomorphosis between body length at sexual maturity and body length at birth in the Cetacea. J. Mammal. Soc. Japan, 3(1): 3–7.
O´Rahilly, R. (1972): Guide to staging of human embryos. Anat. Anz., 130: 556–559.
Perrin, W. F., Coe, J. M., Zweifel, J. R. (1976): Growth and reproduction of the spotted porpoise, *Stenella attenuata,* in the offshore eastern tropical Pacific. Fish Bull., U.S. 75: 725–750.
Perrin, W. F., Reilly, C. B. (1984): Reproductive parameters of dolphins and small whales of the family Delphinidae. Rep. Int. Whal. Comm. (Spec. Issue 6): 97–125.
Pilleri, G., Kraus, C., Gihr, M. (1985): Ontogenesis of *Castor canadensis* in comparison with *Castor fiber* and other rodents. In: Pilleri, G. (ed.): Investigations on beavers. Vol.4: 12–82. (Brain Anatomy Institute, University of Berne).
Schildger, B. (1986): Die externe Morphologie und Wachstumprozesse der drei Delphinarten *Stenella attenuata* (GRAY, 1846), *Stenella longirostris* (GRAY,1828), *Delphinus delphis* Linnaeus,1758 in der pränatalen Entwicklungsphase. Inaug.-Diss. Vet. Med., J.-Liebig – Universität Gießen.
Sedmera, D., Míšek, I., Klima, M. (1997a): On the development of cetacean extremities. I. Hind limb rudimentation in the spotted dolphin *(Stenella attenuata).* European J. Morphol., 35(1): 25–30.
Sedmera, D., Míšek, I., Klima, M. (1997b): On the development of cetacean extremities. II. Morphogenesis and histogenesis of the flippers in the spotted dolphin *(Stenella attenuata).* European J. Morphol., 35(2): 117–123.
Slijper, E. J. (1966): Functional morphology of the reproductive system in Cetacea. In: Norris, K. S. (ed.): Whales, dolphins and porpoises, 277–319. (Univ. Calif. Press, Berkeley – Los Angeles).
Slijper, E. J. (1973): Die Cetaceen – vergleichend-anatomisch und systematisch. (Asher & Co. Amsterdam).
Štěrba, O. (1975): Prenatal growth of the mole, *Talpa europaea* Linn., 1758. Fol. Morphol., 23: 282–285.
Štěrba, O. (1977a): Prenatal development of Central European insectivores. Fol. Zool. 26: 27–44.

Štěrba, O. (1977b): Prenatal development of selected altricial and precocial rodents. Acta Sc. Nat. Brno, 11(11): 1–36.
Štěrba, O. (1979): Prenatal growth of certain artiodactyls. Fol. Zool., 28: 283–288.
Štěrba, O. (1984): Ontogenetic patterns and reproductive strategies in mammals. Fol. Zool., 33: 65–72.
Štěrba, O. (1985): Heterochronies and development in mammals. In: Mlíkovský, J., Novák, V. J. A. (eds): Evolution and morphogenesis, 551–557. (Academia, Praha).
Štěrba, O. (1990): Prenatal development of *Myotis myotis* and *Miniopterus schreibersi*. Fol. Zool., 39: 73–83.
Štěrba, O. (1995): Staging and ageing of mammalian embryos and fetuses. Acta vet. Brno, 64: 83–89.
Štěrba, O., Klima, M., Schildger, B. (1994): Proportional growth of dolphins during prenatal period. Functional Developmental Morphology, 4: 281–283.
Tomilin, A. G. (1940): On biology of the Black Sea dolphin. (K voprosu o biologii chernomorskogo delfina). Tr. Novoross. biol. stancii, 2(3): 31–44. (In Russian).
Tomilin, A. G. (1957): Animals of USSR and of adjacent countries. 9.Cetaceans (Zveri SSSR i prilezhashchih stran. 9. Kitoobraznye). (Publishing House of Academy of Sciences USSR, Moscow, in Russian).
Watson, L. (1981): Sea guide to whales of the world. (Hutchinson, London).
Wells, R. S. (1984): Reproductive behavior and hormone correlates in Hawaiian spinner dolphins, *Stenella longirostris*. Rep. Int. Whal. Comm., Special Issue 6: 465–472.
Yablokov, A. V., Belkovich, V. M., Borisov, V. I. (1972): Whales and dolphins. (Kity i delfiny). (Nauka, Moskva, in Russian).

Subject Index